Grease

Grease

Editors

Raj Shah
Mathias Woydt
Simon C. Tung
Andreas Rosenkranz

MDPI • Basel • Beijing • Wuhan • Barcelona • Belgrade • Manchester • Tokyo • Cluj • Tianjin

Editors

Raj Shah
Koehler Instrument Company
Holtsville, NY 11742, USA

Mathias Woydt
MATRILUB, Materials
Tribology Lubrication,
Limonenstr. 16
D-12203 Berlin-Dahlem,
Germany

Simon C. Tung
Innovation Technology
Consulting Inc.
Glenview, IL 60026, USA

Andreas Rosenkranz
Department of Chemical
Engineering, Biotechnology
and Materials,
Universidad de Chile
Santiago de Chile, Chile

Editorial Office
MDPI
St. Alban-Anlage 66
4052 Basel, Switzerland

This is a reprint of articles from the Special Issue published online in the open access journal *Lubricants* (ISSN 2075-4442) (available at: https://www.mdpi.com/journal/lubricants/special_issues/grease).

For citation purposes, cite each article independently as indicated on the article page online and as indicated below:

LastName, A.A.; LastName, B.B.; LastName, C.C. Article Title. *Journal Name* **Year**, *Volume Number*, Page Range.

ISBN 978-3-0365-3781-8 (Hbk)
ISBN 978-3-0365-3782-5 (PDF)

© 2022 by the authors. Articles in this book are Open Access and distributed under the Creative Commons Attribution (CC BY) license, which allows users to download, copy and build upon published articles, as long as the author and publisher are properly credited, which ensures maximum dissemination and a wider impact of our publications.

The book as a whole is distributed by MDPI under the terms and conditions of the Creative Commons license CC BY-NC-ND.

Contents

About the Editors ix

Raj Shah, Mathias Woydt, Simon C. Tung and Andreas Rosenkranz
Grease
Reprinted from: *Lubricants* **2022**, *10*, 45, doi:10.3390/lubricants10030045 1

Andreas Conrad, Annika Hodapp, Bernhard Hochstein, Norbert Willenbacher and Karl-Heinz Jacob
Low-Temperature Rheology and Thermoanalytical Investigation of Lubricating Greases: Influence of Thickener Type and Concentration on Melting, Crystallization and Glass Transition
Reprinted from: *Lubricants* **2022**, *10*, 1, doi:10.3390/lubricants10010001 5

Adolfo Senatore, Haiping Hong, Veronica D'Urso and Hammad Younes
Tribological Behavior of Novel CNTs-Based Lubricant Grease in Steady-State and Fretting Sliding Conditions
Reprinted from: *Lubricants* **2021**, *9*, 107, doi:10.3390/lubricants9110107 19

Seyedmohammad Vafaei, Dennis Fischer, Max Jopen, Georg Jacobs, Florian König and Ralf Weberskirch
Investigation of Tribological Behavior of Lubricating Greases Composed of Different Bio-Based Polymer Thickeners
Reprinted from: *Lubricants* **2021**, *9*, 80, doi:10.3390/lubricants9080080 31

Daniel Sanchez Garrido, Samuel Leventini and Ashlie Martini
Effect of Temperature and Surface Roughness on the Tribological Behavior of Electric Motor Greases for Hybrid Bearing Materials
Reprinted from: *Lubricants* , *9*, 59, doi:10.3390/lubricants9060059 45

Alan Gurt and Michael M. Khonsari
Testing Grease Consistency
Reprinted from: *Lubricants* **2021**, *9*, 14, doi:10.3390/lubricants9020014 61

Michael M. Khonsari, K. P. Lijesh, Roger A. Miller and Raj Shah
Evaluating Grease Degradation through Contact Angle Approach
Reprinted from: *Lubricants* **2021**, *9*, 11, doi:10.3390/lubricants9010011 77

Kunihiko Kakoi
Formulation to Calculate Isothermal, Non-Newtonian Elastohydrodynamic Lubrication Problems Using a Pressure Gradient Coordinate System and Its Verification by an Experimental Grease
Reprinted from: *Lubricants* **2021**, *9*, 56, doi:10.3390/lubricants9050056 91

Raj Shah, Simon Tung, Rui Chen and Roger Miller
Grease Performance Requirements and Future Perspectives for Electric and Hybrid Vehicle Applications
Reprinted from: *Lubricants* **2021**, *9*, 40, doi:10.3390/lubricants9040040 105

Sravan K. Joysula, Anshuman Dube, Debdutt Patro and Deepak Halenahally Veeregowda
On the Fictitious Grease Lubrication Performance in a Four-Ball Tester
Reprinted from: *Lubricants* **2021**, *9*, 33, doi:10.3390/lubricants9030033 121

Emmanuel P. Georgiou, Dirk Drees, Michel De Bilde, Michael Anderson, Matthias Carlstedt and Olaf Mollenhauer
Quantification of Tackiness of a Grease: The Road to a Method
Reprinted from: *Lubricants* **2021**, *9*, 32, doi:10.3390/lubricants9030032 . . . . **133**

About the Editors

Raj Shah is a Director at Koehler Instrument Company, in Long Island, NY where he has been working for over 25 years. He holds a Ph. D in Chemical Engineering from The Pennsylvania State University and Fellowship from the Chartered Management Institute, London. Dr. Shah is an elected Fellow by his peers at STLE, NLGI, AIC, RSC, INSTMC, IChemE, CMI, IOP and EI. He also holds all six of these highly coveted certifications: namely CPC, CCHE, CEng, CSci, CChem, CPEng. Dr. Shah was on the Board of Directors for 15 years of National Lubricating Grease Institute, and vice chair of ASTM D02. G (Grease committee) for over 10 years and is the co-editor of the fully revised 2022 NLGI Grease handbook. He is the recipient of numerous awards from ASTM, STLE and NLGI, Dr. Shah is also an adjunct Professor in the Dept of Material Science and Chemical Engineering at State University of New York, Stony Brook and serves on the advisory board of directors of several professional societies and universities in the USA. A Tau Beta Pi eminent engineer, he has over 475 publications and lives in Long Island, NY.

Mathias Woydt is managing partner of MATRILUB Materials I Tribology I Lubrication, with more than 36 years of experience in R&D and with more than 360 publications and 51 priority patents filed. He is a STLE Fellow, recipient of an ASTM award of excellence and board member of the German Society for Tribology.

Simon C. Tung holds a PhD in Chemical Engineering from Rensselaer Polytechnic Institute and an MBA Degree from University of Michigan-Ann Arbor. Dr. Tung has been involved in the automotive industry since joining the Fuels and Lubricants (F&L) Department at General Motors Research Laboratories in 1982. While there, Dr. Tung led pioneering research and development on automotive propulsion systems, powertrain tribology, fuels, and lubricants (F&L) performance testing, and advanced manufacturing lubrication processes going back approximately 30 years. During the period 2000-2008, Dr. Tung was appointed as a Technical Fellow and Engineering Group Manager to lead the advanced energy efficient propulsion technology and discovery research programs in the General Motors R&D Center. Dr. Tung led GM R&D and manufacturing teams in improving powertrain energy efficiency and drivetrain durability.

Dr. Tung has been engaged in various leadership role working with automotive and chemical industries. During that time, he worked at General Motors R&D Center, Vanderbilt, and Industrial Technology Research Institute (ITRI). Dr. Tung was responsible to lead R&D teams for improving energy efficiency capability in advanced powertrain or manufacturing lubrication systems. Dr. Tung has accomplished many outstanding R&D programs for powertrain/hybrid propulsion vehicles, powertrain/manufacturing lubrication, tribology testing, thermal management design, and global OEM lubricant liaison program management.

In 2008, Dr. Tung joined the Industrial Technology Research Institute (ITRI) as General Director, where Dr. Tung was responsible to manage all R&D programs in the Energy and Environmental Research Laboratories. Dr. Tung made significant contributions in the research areas of green energy, energy storage system, fuel cell, hydrogen energy, and reducing greenhouse emissions. Dr. Tung's technical expertise includes powertrain engineering, tribology control, energy storage materials, green energy technology, and advanced propulsion systems.

At present, Dr. Tung is serving as the Senior Consultant to coach OEMs in the areas of advanced lubricants R&D and manufacturing lubrication at Innovation Technology Consulting. Dr. Tung has secured 25 US/International patents, most of which have been integrated into modern automotive functionalities. Dr. Tung has published 195 papers and presentations, 5 reference handbooks, 165 journal papers, and 25 U.S. and international patents. Dr. Tung has received 26 outstanding achievement awards and delivered 25 invited keynote presentations,

Dr. Tung has received many distinguished honors, including being named as an STLE Fellow in the Society of Tribologists and Lubrication Engineers (STLE), SAE Fellow, as well as receiving the Gold Award from the Engineering Society of Detroit (ESD). Dr. Tung was honored with SAE International's Edward N. Cole Award for Automotive Engineering Innovation during the SAE 2011 World Congress. In addition, Dr. Tung has been recognized by SAE as the winner of the SAE Franz Pischinger Powertrain Innovation Award in April 2013, based on his outstanding professional achievements, innovative research, and his leadership in the fields of automotive engineering, and powertrain systems. Dr. Tung also won the SAE International Leadership Citation Award during the SAE 2015 World Congress to recognize his distinguished leadership and international contributions. Dr. Tung is an internationally recognized expert on automotive powertrain tribology, advanced materials implementation, driveline lubrication and the commercialization of innovation products in propulsion technology, including tribology control, energy storage materials, and thermal management technologies.

Andreas Rosenkranz is a Professor for Materials-Oriented Tribology and New 2D Materials in the Department of Chemical Engineering, Biotechnology and Materials at the University of Chile. His research focuses on the characterization, chemical functionalization, and application of new 2D materials. His main field of research is related to tribology (friction, wear, and energy efficiency), but in the last couple of years, he has also expanded his fields towards water purification, catalysis, and biological properties. He has published more than 100 peer-reviewed journal publications, is a fellow of the Alexander von Humboldt Foundation and acts as a scientific editor for different well-reputed scientific journals including Applied Nanoscience and Frontiers in Chemistry.

Editorial

Grease

Raj Shah [1,*], Mathias Woydt [2], Simon C. Tung [3] and Andreas Rosenkranz [4]

1. Koehler Instrument Company, Holtsville, NY 11742, USA
2. MATRILUB, Limonenstr. 16, D-12203 Berlin-Dahlem, Germany; m.woydt@matrilub.de
3. Innovation Technology Consulting Inc., 1801 Tower Drive, E203, Glenview, IL 60026, USA; simontung168@gmail.com
4. Department of Chemical Engineering, Biotechnology and Materials, Universidad de Chile, Santiago de Chile 8450000, Chile; arosenkranz@ing.uchile.cl
* Correspondence: rshah@koehlerinstrument.com

Grease is an extraordinarily complex lubricant with a complex material–property relationship, and to shed more light on its importance, we decided to launch the first Special Issue of "*Lubricants*" purely focusing on the most recent developmental trends of grease applications. In recent years, significant research progress has been achieved concerning greases applied for electrified powertrains, ranging from specific grease chemical formulations for special applications to how grease interacts with various surfaces. This Special Issue has compiled top research papers related to lubricating greases, hoping to establish an annual trend for discussing the latest global developments encompassing all R&D areas related to the innovation development of greases for automotive and manufacturing industries.

In total, six research articles, two technical notes, one perspective, and one case report were compiled, all pertaining to tribology and lubricating greases. Conrad et al. investigated the influence of thickeners on the crystallization, melting, and glass transition of lubricant greases. The type and concentration of the added thickeners had a notable impact on the properties of greases formulated from mineral oil, polyalphaolefin, alkylated naphthalene, propylene glycol, and trimellitate [1]. Three manuscripts focus on analyzing the tribological behavior and performance of varying greases for their intended applications. Senatore et al. studied the tribological behavior of novel 7.5 wt.-% carbon nanotube-based (CNT) lubricant greases in polyalphaolefin (PAO) with and without 1.0 wt.-% MoS_2. Their results indicated that the novel CNT-based greases exhibited superior tribological properties when compared against other commercial greases [2].

Meanwhile, Vafaei et al. compared and evaluated the lubrication properties of three different bio-based polymer thickener systems and developed bio-based greases via a ball-on-disc tribometer [3]. In a study presented by Garrido et al., the tribological performance of four commercial electric motor (EM) greases, with varying quantities of polyurea or lithium thickener with mineral or synthetic-based oil, were evaluated through the measurement of friction and the wear of silicon nitride sliding on hardened 52,100 bearing steel. The results provided an explicit comparison of commercially available EM greases across a wide range of applications and relevant metrics [4].

Gurt and Khonsari highlighted the relevant parameters associated with the rheometer penetration test and the recommended testing procedure for measuring the consistency of various greases. Their results were compared to data obtained from yield stress, crossover stress, and cone penetration tests [5]. For a different methodology, Khonsari et al. detailed the results of a novel approach for the evaluation of the water-resistance of greases to quantify degradation. This newly developed approach, known as the contact angle approach, involves the measurement of the contact angle of a water droplet on the surface of a sample of grease [6].

Kakoi presented a formulation of a point-contact elasto-hydrodynamic lubrication analysis for an isothermal, non-Newtonian flow, with the employment of a coordinate system of pressure gradient. The formulation detailed in this study was applied to a grease that had previously been evaluated to compare results to verify the validity of the formulation [7].

On a different subject for grease application in EV and hybrid vehicles (HVs), Shah et al. [8] discussed the role of grease lubrication in electric vehicles (EVs) and hybrid vehicles (HVs) in terms of performance requirements. The future development of lubricating grease used in EVs and HVs needs to be improved for meeting the lubrication and thermal management requirements [8]. Shah et al. also pointed out greases need to be formulated for new factors in electrical vehicles (EVs), including the increased presence of electricity, electrical currents, and noise in an EV due to the absence of an internal combustion engine (ICE) [8]. The major differences between EVs and conventional ICEVs can be grouped into the following technical areas: energy efficiency, noise, vibration, harshness (NVH) issues, the presence of an electrical current and electromagnetic fields from electric modules, sensors, and circuits, and bearing lubrication. Additional considerations include the thermal transfer, seals, corrosion protection, and materials' compatibility. Shah et al. reviewed the future development trends of EVs/HVs on driveline lubrication and thermal management requirements. Due to the increased number of electrical components, such as electric modules and sensors, greases must be formulated to be unreactive with electricity. In addition, the role of grease lubrication in electric vehicles (EVs) and hybrid vehicles (HVs) is crucial in terms of performance requirements. Comparisons of grease lubrication in EVs and HVs from IC engines for performance requirements were reviewed in terms of electrical and thermal properties under different operating conditions.

Loysula et al. investigated the fictitious lubrication performance in a four-ball tester in accordance with ASTM D2596. The findings of this study indicated that the parameter "speed ramp up time" is an essential component that should be researched by grease manufacturers to prevent the use of grease with fictitious extreme pressure (EP) behavior [9]. Georgiou et al. highlighted the development of a reliable, quantitative method for measuring the tackiness and adhesion of greases. The study highlighted the influence of temperature on the tackiness of greases and the reproducibility of the standardized tackiness method [10].

The editors of this Special Issue would like to provide a final conclusion regarding the importance of grease for future EV/hybrid vehicle lubrication applications [11,12]. The future development of electric vehicles will globally influence the selection and development of gear oils, coolants, and greases, as they will be in contact with electric modules, sensors, and circuits, and will be affected by the electrical current and electromagnetic fields. The increasing presence of electrical parts in EVs/HVs requires the corrosion protection of bearings and other remaining mechanical components. Thus, it is imperative for specialized greases to be explored for specific applications in EVs/HVs to ensure maximum protection from friction, wear, and corrosion to guarantee the longevity of the operating automobile [11,12].

Author Contributions: Conceptualization, methodology, validation, writing—original draft preparation, writing—review and editing, all authors together. All authors have read and agreed to the published version of the manuscript.

Funding: This research received no external funding.

Institutional Review Board Statement: Not applicable.

Informed Consent Statement: Not applicable.

Data Availability Statement: Not applicable.

Conflicts of Interest: The authors declare no conflict of interest.

References

1. Conrad, A.; Hodapp, A.; Hochstein, B.; Willenbacher, N.; Jacob, K.-H. Low-Temperature Rheology and Thermoanalytical Investigation of Lubricating Greases: Influence of Thickener Type and Concentration on Melting, Crystallization and Glass Transition. *Lubricants* **2022**, *10*, 1. [CrossRef]
2. Senatore, A.; Hong, H.; D'Urso, V.; Younes, H. Tribological Behavior of Novel CNTs-Based Lubricant Grease in Steady-State and Fretting Sliding Conditions. *Lubricants* **2021**, *9*, 107. [CrossRef]
3. Vafaei, S.; Fischer, D.; Jopen, M.; Jacobs, G.; König, F.; Weberskirch, R. Investigation of Tribological Behavior of Lubricating Greases Composed of Different Bio-Based Polymer Thickeners. *Lubricants* **2021**, *9*, 80. [CrossRef]
4. Sanchez Garrido, D.; Leventini, S.; Martini, A. Effect of Temperature and Surface Roughness on the Tribological Behavior of Electric Motor Greases for Hybrid Bearing Materials. *Lubricants* **2021**, *9*, 59. [CrossRef]
5. Gurt, A.; Khonsari, M.M. Testing Grease Consistency. *Lubricants* **2021**, *9*, 14. [CrossRef]
6. Khonsari, M.M.; Lijesh, K.P.; Miller, R.A.; Shah, R. Evaluating Grease Degradation through Contact Angle Approach. *Lubricants* **2021**, *9*, 11. [CrossRef]
7. Kakoi, K. Formulation to Calculate Isothermal, Non-Newtonian Elastohydrodynamic Lubrication Problems Using a Pressure Gradient Coordinate System and Its Verification by an Experimental Grease. *Lubricants* **2021**, *9*, 56. [CrossRef]
8. Shah, R.; Tung, S.; Chen, R.; Miller, R. Grease Performance Requirements and Future Perspectives for Electric and Hybrid Vehicle Applications. *Lubricants* **2021**, *9*, 40. [CrossRef]
9. Loysula, S.K.; Dube, A.; Patro, D.; Veeregowda, D.H. On the Fictitious Grease Lubrication Performance in a Four-Ball Tester. *Lubricants* **2021**, *9*, 33. [CrossRef]
10. Georgiou, E.P.; Drees, D.; De Bilde, M.; Anderson, M.; Carlstedt, M.; Mollenhauer, O. Quantification of Tackiness of a Grease: The Road to a Method. *Lubricants* **2021**, *9*, 32. [CrossRef]
11. Shah, R.; Gashi, B.; González-Poggini, S.; Colet-Lagrille, M.; Rosenkranz, A. Recent trends in batteries and lubricants for electric vehicles. *Adv. Mech. Eng.* **2021**, *13*, 16878140211021730. [CrossRef]
12. Shah, R.; Mittal, V.; Matsil, E.; Rosenkranz, A. Magnesium-ion batteries for electric vehicles: Current trends and future perspectives. *Adv. Mech. Eng.* **2021**, *13*, 16878140211003398. [CrossRef]

Article

Low-Temperature Rheology and Thermoanalytical Investigation of Lubricating Greases: Influence of Thickener Type and Concentration on Melting, Crystallization and Glass Transition

Andreas Conrad [1,*], Annika Hodapp [2], Bernhard Hochstein [2], Norbert Willenbacher [2] and Karl-Heinz Jacob [1]

1. Applied Chemistry, Nuremberg Institute of Technology Georg Simon Ohm, 90489 Nuremberg, Germany; karl-heinz.jacob@th-nuernberg.de
2. Institute of Mechanical Process Engineering and Mechanics, Karlsruhe Institute of Technology, 76131 Karlsruhe, Germany; annika.hodapp@kit.edu (A.H.); bernhard.hochstein@kit.edu (B.H.); norbert.willenbacher@kit.edu (N.W.)
* Correspondence: andreas.conrad@th-nuernberg.de

Abstract: This study investigates crystallization, melting and glass transition of Li- and Ca-12-hydroxystearate greases in relation to the pour point of the corresponding oils. The base oils for the greases are mineral oil, polyalphaolefin, alkylated naphthalene, propylene glycol, and trimellitate. For the mineral oil-based greases the crystallization temperature T_c increases and the melting temperature T_m decreases upon addition of thickener. The pour point of the mineral oil then is 3 K below T_c and does not properly define the lowest application temperature for mineral oil (MO) based greases. Both thickeners induce a small increase of the glass transition temperature (1–3 K) of the synthetic oils polyalphaolefin, alkylated naphthalene, propylene glycol. The pour point of the base oils correlates well with the onset of the glass transition in the corresponding grease indicated by a sharp increase in grease viscosity. Pure trimellitate with unbranched alkyl chains does not crystallize upon cooling but shows noticeable supercooling and cold crystallization. As the percentage of thickener in corresponding greases increases, more oil crystallizes upon cooling 20 K above the crystallization temperature of the trimellitate without thickener (−44 °C). Here, the thickener changes the crystallization behavior from homogeneous to heterogeneous and thus acts as a crystallization nucleus. The pour point of the base oil does not provide information on the temperature below which the greases stiffen significantly due to crystallization.

Keywords: lubricating grease; heterogeneous crystallization; glass transition; rheology; differential scanning calorimetry (DSC)

1. Introduction

Concerning the lubrication conditions in operation, the higher viscosity of the base oils and higher consistency of the greases at low temperatures, the friction factor increases slightly [1]. In the case of mineral oils, as the proportion of paraffin crystals increases with decreasing temperature, the sliding friction also increases. Nonetheless, the outflow of mineral oil provides adequate lubrication during sliding [2]. Unlike mineral oil, ester oils crystallize to the extent of blocking the rheometer [3]. Thus, it would also block a tribological contact during operation. Practical test methods such as the low-temperature torque test for wheel bearings (ASTM D4693 [4]) or ball bearings (ASTM D1478 [5]) determine the suitability of greases for low temperatures. Although these standards are close to practical experience, they do not provide information on whether the base oil in lubricating greases precipitate crystals and thus change their flow properties upon cooling. In practice, the pour point according to ASTM D97 [6] is measured for this purpose. The pour point indicates the temperature when the base oil stops flowing as a sample vessel is tilted. Previous research has shown that crystallization and viscosity increase are responsible for oils stop flowing at temperatures below the pour point. However, some oils can crystallize at temperatures above the pour point if given enough time [3].

Mineral oils precipitate paraffin crystals below the crystallization temperature (T_c), turning Newtonian mineral oils into shear-thinning suspensions. The transition from Newtonian to viscoelastic behavior is indicated by a significant slope increase in an Arrhenius diagram ($\ln(\eta)$-T^{-1}) below the crystallization temperature [7]. The necessary supercooling ($\Delta T = T_c - T_m$) for precipitation of paraffin crystals is hardly dependent on the shear rate and the cooling conditions and is almost constant at about $\Delta T = -10$ K. The crystallization temperature can be equated with the pour point of mineral oils [3].

Unlike mineral oils, which are mixtures of paraffinic, naphthenic, and aromatic hydrocarbons, synthetic oils consist of chemically uniform compounds with a comparatively narrow molecular weight distribution. The synthetic oils polyalphaolefin, polyalkylenglycole, alkylated naphthalene, and tris-(2-ethylhexyl)trimelltiate solidify glass-like below -70 °C. Upon cooling, the viscosity of these base oils increases steadily and follows a Williams–Landes–Ferry Equation (WLF) down to 20 K above the glass transition temperature [3].

Ester oils with linear alkyl chains and a narrow molecular weight distribution crystallize with strong supercooling effects [8]. Using the example of a trimellitate with linear alkyl chains (C8–C10), the viscosity increases steadily up to the crystallization temperature. At the crystallization temperature, the viscosity rises abruptly. However, crystallization does not lead to shear-thinning suspensions as in mineral oils but to a solid [3].

The crystallization temperatures of a base oil in a lubricating grease may differ from the pure base oil due to the catalytic effect the thickener on crystallization, i.e., heterogeneous crystallization. From a thermodynamic point of view, nucleation in a base oil is spontaneous when the size of nuclei corresponds to a critical size, which decreases with increasing supercooling ($\Delta T = T_c - T_m$). Particles in a liquid can act as nuclei with a larger critical nucleation radius, resulting in a lower energy barrier for nucleation and less supercooling [9]. The necessary condition for particles to catalyze nucleation is that they have melting temperatures far higher than the melting point of the lubricating oil and remain solid, which is the case for common thickeners such as Li- and Ca-12-hydroxystearate [10].

From a kinetic point of view, the rate of nucleus formation depends on viscosity, i.e., the lower the temperature, the lower the nucleation rate. Approaching the glass-transition temperature T_G, the nucleation rate becomes infinitely small. Since the temperatures for a maximum nucleation rate and maximum crystal growth rate are not identical, lubricating oil composition and cooling rate significantly influence supercooling. When the maximum crystal growth rate temperature is below the glass temperature T_G, nuclei are absent, and glass-like solidification takes place upon cooling [10,11].

If such base oils are processed into lubricating greases with Li- or Ca-12-hydroxy stearate, the thickener type and concentration determine the consistency of the metal soap greases. The metal soap must be melted in the base oil and then cooled under defined conditions. During this process, a thickener structure forms, which is responsible for the viscoelasticity of the lubricating greases [12]. With Li-12-hydroystearate as a thickener, a network of platelets is formed at low concentrations, which changes into fine and dense fibril-like structures at higher concentrations [13]. Calcium complex soaps, for example, build globular structures [14]. Even at relatively low thickener concentrations, metallic soaps cause viscoelasticity of the lubricating greases and, above a specific concentration, forming a viscoelastic liquid with a yield point [15].

In the context of base oils and associated lubricating greases, the question arises whether and what influence the thickener type, and concentration have on crystallization, melting, and glass transition temperatures. These questions are discussed in detail below, based on rheological and thermoanalytical measurements for various lubricating greases. The relevance of the base oils' pour point for the low-temperature behavior of corresponding greases will also be addressed.

2. Materials and Methods

Table 1 lists the base oils used with the kinematic viscosities at 40 and 100 °C, the viscosity index, and the respective pour point. The crystallization behavior and low-temperature rheology is previously examined in detail and the base oils are classified in three groups [3]. Group I contains a mineral oil (MO), Group II includes amorphously solidifying synthetic lubricating oils (PAG, KR-008, PAO8), and Group III comprises a crystallizing synthetic lubricating oil (EO).

Table 1. Classification (Group I–III), kinematic viscosity (ν), viscosity index (VI), pour point (ASTM D7346 [16]), and chemical nature of the base oils.

	Chemical Nature	Group	ν @40 °C/cSt.	ν @100 °C/cSt.	VI	Pour Point/°C
MO	mineral oil (SN 100/SN 500)	I	48	5.3	105	−12
PAO8	polyalphaolefin	II	47	8	139	−66
KR−008	alkylated naphthalene	II	36	7	68	−54
PAG	polypropylenglycole	II	57	10.4	188	−51
EO	Trimellititate *	III	52	8.1	128	−57

* with linear C8–C10 alkyl groups.

From the base oils listed in Table 1, lubricating greases with Li- and Ca-12-hydroxystearate were prepared. Ca-12-hydroxystearate greases were prepared by melting the Ca-12-hydroxystearate in the base oil at 120 °C for 30 min, while the Li-12-hydroxystearate lubricating greases were prepared by melting Li-12-hydroxystearate above the melting point of 212 °C. After cooling to room temperature, homogenization of the cooled suspensions was carried out on a three-roll mill (Exakt Advanced Technologies GmbH, 50I, Norderstedt, Germany). Greases with thickener concentrations lower than 5 wt.% were homogenized with an Ultra-Turrax (IKA GmbH, Staufen, Germany). Table A1 in the appendix lists the exact composition of the lubricating greases with the respective worked penetration P_w (DIN ISO 2137 [17]) and corresponding NLGI class (DIN 51818 [18]).

2.1. Rheological Measurements

The steady shear measurements were performed with MCR301 and 702 rheometers from Anton Paar (Ostfildern-Scharnhausen, Germany). A plate-plate geometry with 25 mm in diameter (PP25) made of stainless steel was used as measuring system with a gap of 1 mm. In the temperature range between 20 and −40 °C the Peltier unit was purged with dry air (dew point: −80 °C) to prevent condensation and freezing of humidity. Measurements below −40 °C were performed using a PP25 geometry covered by a low-temperature CTD450 cell and an EVU10 controller for liquid nitrogen, both from Anton Paar (Ostfildern-Scharnhausen, Germany). Temperature-dependent oscillatory shear measurements were performed at an angular frequency of $\omega = 10$ rad s^{-1} and an amplitude of $\gamma = 0.05\%$. Strict care was taken to ensure that the linear viscoelastic (LVE) range was maintained in the temperature range investigated. All rheological experiments were performed in triplicate with fresh samples for each measurement.

2.2. Differential Scanning Calorimetry (DSC)

Heat flow measurements during the cooling and heating cycles were performed with a DSC 204 F1 Phoenix® (Netzsch, Selb, Germany) in a pierced aluminum pan with a sample weight of approx. 10 mg to detect glass transition, crystallization, and melting of the lubricating greases. The measurements were carried out in the temperature range from 25 to −60 °C with heating and cooling rate of 2 K min^{-1}. For the extended temperature range down to −180 °C, a Netzsch DSC 204 cell with a CC 200 L controller for liquid nitrogen was used. The lubricating grease samples were cooled with −20 K min^{-1} from 25 °C to −180 °C, held for 5 min, and then heated with a rate of +10 K min^{-1}. All thermoanalytical experiments were performed in triplicate with fresh samples for each measurement.

3. Results and Discussion

3.1. Crystallization and Melting of Lubricating Greases Based on Mineral Oil (Group I)

Figure 1 shows the absolute value of the complex viscosity $|\eta^*|$ as a function of the temperature for the pure mineral oil (MO) and two mineral oil greases, with different Li- (a) and Ca-12-hydroxystearate (b) concentrations during cooling (empty symbols) and heating (filled symbols). Upon cooling, the complex viscosity increases steadily until the crystallization temperature $T_{c,rheo}$ is reached. Below $T_{c,rheo}$, the slope of the complex viscosity curve increases sharply. The complex viscosity increases levels off below $-20\,°C$. On reheating, the complex viscosity of the greases first decreases sharply until the melting temperature $T_{m,rheo}$ is reached, which is always higher than $T_{c,rheo}$. Above $T_{m,rheo}$, the absolute values of the complex viscosities of the cooling and heating cycles overlap. The reason for the sharp change in slopes on cooling and heating, respectively, is the formation and dissolution of paraffin crystals. Previous investigations on the mineral oil MO have shown that the MO changes from a Newtonian liquid to a shear-thinning suspension below $T_{c,rheo}$ [3].

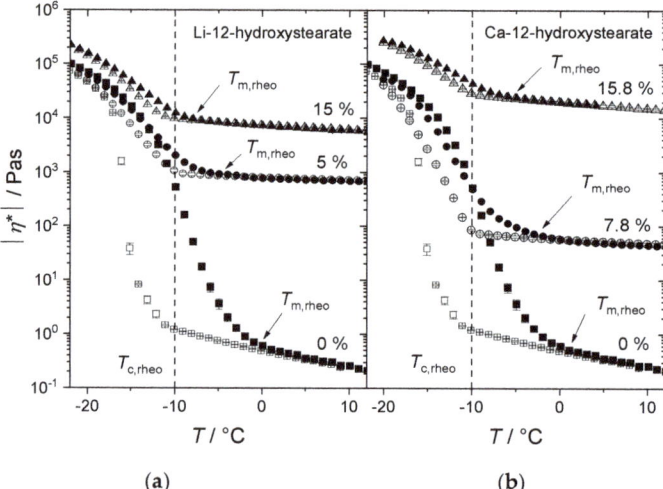

Figure 1. Absolute value of the complex viscosity $|\eta^*|$ as a function of temperature from cooling (empty) and heating (filled) of pure mineral oil and two lubricating greases based on mineral oil of different proportions of (a) Li and (b) Ca-12 hydroxystearate at a heating and cooling rate β of 2 K min^{-1}, angular frequency ω of 10 rad s^{-1} and deformation γ of 0.05%. The dashed line marks the crystallization temperature $T_{c,rheo}$ and the melting temperatures $T_{m,rheo}$ are marked by arrows.

Above the melting temperature $T_{m,rheo}$, the temperature range in which the oil behaves Newtonian, the level of the complex viscosity increases with increasing thickener concentration from 0.5 Pas without thickener to approx. 10^4 Pas with 15.0 wt.% Li-12-hydroxystearate and 15.8 wt% Ca-12-hydroxystearate as thickener, respectively. Below the temperature of $-20\,°C$, the viscosity reaches a level of about 10^5 Pas regardless of thickener type and concentration. The complex viscosity of the greases at temperatures lower than $-20\,°C$ is affected mainly by the presence of paraffin crystals.

Figure 2 depicts the crystallization temperatures $T_{c,rheo}$ and melting temperatures $T_{m,rheo}$ of mineral oil greases with Li- (a) and Ca-12-hydroxystearate (b) as a function of thickener concentration. $T_{c,rheo}$ and $T_{m,rheo}$ were obtained by intersecting two tangents above and below the significant slope change in Figure 1. The mineral oil without thickener exhibits a crystallization temperature $T_{c,rheo}$ of $-11.7 \pm 0.15\,°C$ and a melting temperature $T_{m,rheo}$ of $-0.8 \pm 1\,°C$. Regardless of Li- or Ca-12-hydroxystearate greases, the crystallization temperatures increase to values between -8 and $-10\,°C$, and the melting temperatures decrease to values between −5 and −7 °C. The supercooling ($\Delta T = T_c - T_m$) decreases from approx. 11 K without thickener to approx. 3 K for Li-12-hydroxystearate and approx. 5 K for Ca-12-hydroxystearate greases at a thickener concentration of about 5 wt.% and essentially remains constant when the fraction of thickener is further increased.

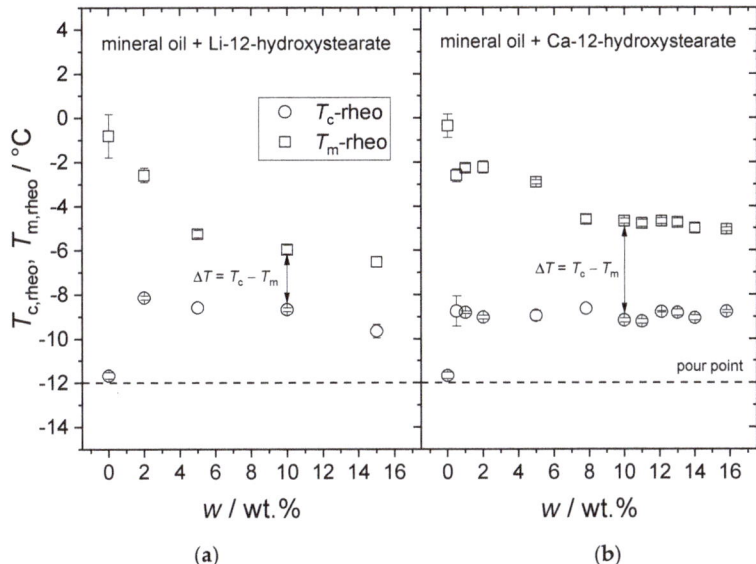

Figure 2. Crystallization- ($T_{c,rheo}$) and melting temperature ($T_{m,rheo}$) as a function of thickener concentration w for Li-12-hydroxystearate (**a**) and Ca-12-hydroxystearate (**b**) based on mineral oil (MO), obtained from small amplitude oscillatory shear rheometry cooling and heating cycles (see Figure 1) with an angular frequency $\omega = 10$ rad s^{-1}, deformation $\gamma = 0.05\%$ cooling and heating rate of $\beta = 2$ K min^{-1}. $T_{c,rheo}$, and $T_{m,rheo}$ are obtained by intercepting two tangents above and below the significant slope change in Figure 1. The supercooling ($\Delta T = T_c - T_m$) decreases from approx. 11 K without thickener to approx. 3 K with Li-12-hydroxystearate and approx. 5 K with Ca-12-hydroxystearate. A dashed line marks the pour point of −12 °C.

The increase in crystallization temperature $T_{c,rheo}$ is presumably caused by the thickener particles in the base oil. The thickener particles provide crystal nuclei, favoring the formation of paraffin crystals at higher temperatures, here about −8 °C [9,10]. Consequently, the crystallization temperature of the lubricating greases is 4 K above the pour point of the mineral oil (−12 °C). A small amount of dissolved thickener in the base oil presumably causes the decrease in T_m [19].

3.2. Lubricating Greases Based on Non-Crystallizing Synthetic Base Oils (Group II)

Figure 3a displays the absolute value of the complex viscosity and Figure 3b the specific heat flow $\dot{q}$ as a function of temperature for the grease with the synthetic base oil PAO8 and 22 wt.% Li-12-hydroxystearate (PAO8-22) as a thickener. In the logarithmic plot, the complex viscosity increases approximately linearly with decreasing temperature up to the onset of the glass transition ($T_{G,PAO8-22} = -83 \pm 0.2$ °C). The absolute value of the complex viscosity increases sharply in the temperature interval between −66 °C and T_G, but remains constant at a high level of 10^7 Pas below T_G. The lubricating greases based on KR-008 (Figure A1) and PAG (Figure A2) behave similarly, see Appendix A. Notably, the temperature at which the two tangents fitted to the sections of the $|\eta^*|$ curve with different slope agrees with the pour point and the end of the section in which $|\eta^*|$ steeply increases corresponds to the glass transition temperature. According to the nucleation theory, for oils such as PAO8, the maximum temperature of crystal growth rate is below the glass temperature T_G [10,11].

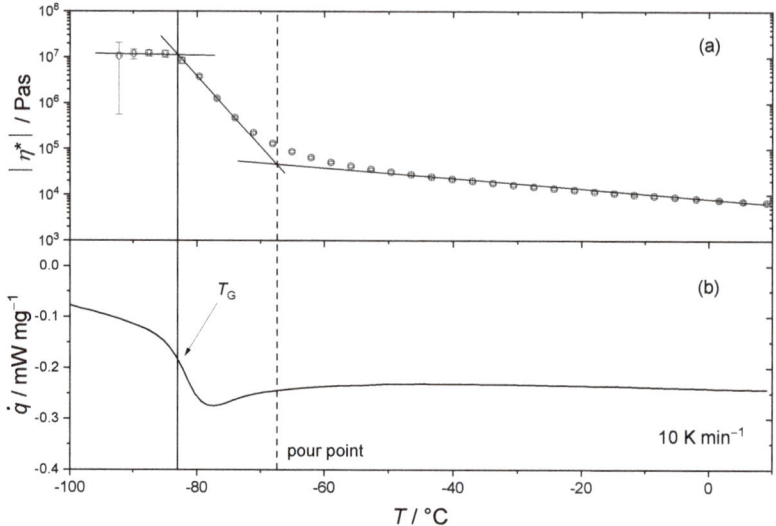

Figure 3. (a) Absolute value of complex viscosity $|\eta^*|$ as a function of temperature for the lubricating grease made from the base oil PAO8 with 22 wt.% Li-12-hydroxystearate as a thickener (PAO8-22), measured in oscillatory shear with the deformation amplitude $\gamma = 0.05\%$, angular frequency $\omega = 10$ rad s^{-1} and cooling rate of -10 K min^{-1}. Lines are to guide the eye. (b) Specific heat flow $\dot{q}$ of the PAO8-22 grease at a heating rate of 10 K min^{-1}. PAO8 shows a glass transition temperature T_G at -83 ± 0.2 °C, marked by a solid line. The dashed line marks the pour point for the base oil PAO8 of -66 °C.

Figure 4 compares the glass transition temperatures T_G of the base oils PAO8, KR-008 and PAG (abscissa) and the corresponding lubricating greases (ordinate) with Li and Ca-12-hydroxystearate as a thickener. Glass transition temperature of the lubricating greases is always slightly higher (≈ 2 K) than that of the corresponding base oil but hardly depends on the thickener concentration at least in the investigated concentration range and stays always below -70 °C. The solidification behavior of greases based on non-crystallizing synthetic base oils does not change significantly due to the added thickener. The pour point as well as the glass transition temperature for these base oils and lubricating greases, respectively, are good indicators for their lowest application temperature.

3.3. Greases Based on Crystallizing Synthetic Base Oils (Group III)

The absolute value of the complex viscosity of the lubricating greases based on the trimellitate EO increases uniformly on cooling down to the crystallization temperature T_c and then rises abruptly. The ester crystallizes to such an extent that the rheometer blocks, and no further measurements are possible [3]. For this reason, the crystallization behavior of trimellitate greases was investigated primarily using differential scanning calorimetry (DSC).

Figure 5 shows the heat flux from DSC measurements during cooling (a) and reheating (b) for Ca-12 hydroxystearate lubricating greases with thickener concentrations between 0.5 and 13 wt.%. During cooling, in the temperature range between -10 °C and -50 °C, an exothermic signal arises at thickener concentrations higher than 1 wt.%, which increases in magnitude with increasing thickener concentration and the peak shifts to higher temperatures. The onset temperature increases slightly.

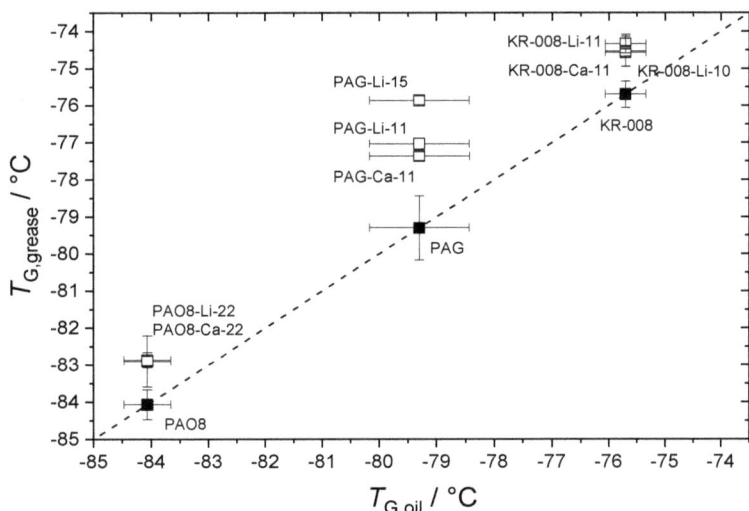

Figure 4. Comparison of the glass transition temperatures of pure base oils PAO8, KR-008, and PAG (abscissa) and lubricating greases with different Li and Ca-12-hydroxystearate concentrations (ordinate) obtained from DSC-measurements at 10 K min^{-1}. PAO8 with 22 wt.% (PAO8-Li-22, PAO8-Ca-22), KR-008 with 10 and 11 wt.% (KR-008-Li-10, KR-008-Li-11, KR-008-Ca-11) and PAG with 11 and 15 wt.% (PAG-Li-11, PAG-Li-15, PAG-Ca-11). The dashed line represents the angle bisector on which the T_G's of the pure oils lie (PAO8, PAG, and KR-008).

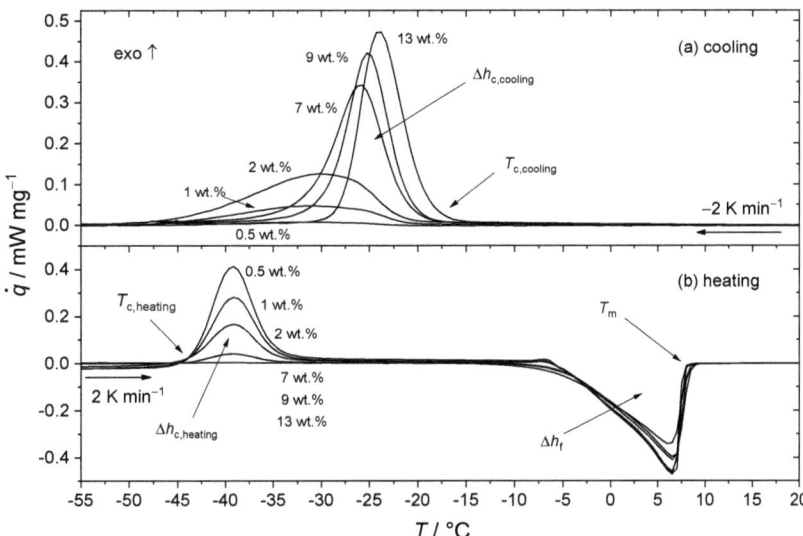

Figure 5. Heat flow $\dot{q}$ as obtained from DSC measurements during cooling (**a**) and reheating (**b**) for Ca-12 hydroxystearate greases with thickener concentrations of 0.5, 1, 2, 7, 9, and 13 wt.% with trimellitate (EO) as base oil at a cooling (**a**) and heating (**b**) rate of 2 K min^{-1}. The integral of the exothermic signals during cooling and heating represent the crystallization enthalpy $\Delta h_{c,cooling}$ and $\Delta h_{c,heating}$, respectively. The integral of the endothermic peak represents the enthalpy of fusion (i.e., melting) Δh_f. $T_{c,cooling}$ and $T_{c,heating}$ are the onset temperatures of the crystallization peaks. T_m is the offset of the melting peak.

In addition, an exothermic peak is observed during heating in the temperature range between −45 °C and −30 °C. The crystallization enthalpy during heating $\Delta h_{c,\text{heating}}$ decreases with increasing thickener concentration and is not measurable anymore for thickener concentrations >9 wt.%; peak temperatures and peak widths remain the same. Above −10 °C, melting of the sample begins. Melting temperature and area of the endothermic peak do not change with thickener concentration. However, the peak becomes somewhat flatter and broader with increasing thickener concentration.

Figure 6 shows melting temperature T_m and crystallization temperature during cooling ($T_{c,\text{cooling}}$) and heating ($T_{c,\text{heating}}$) of the trimellitate (EO) based Li- and Ca-12-hydroxystearate greases as a function of thickener content. The melting temperature $T_m = 7.8 \pm 0.4$ °C is independent of thickener type and concentration and the same is true for the crystallization temperature determined during the heating cycle $T_{c,\text{heating}} = -46.1 \pm 1.9$ °C.

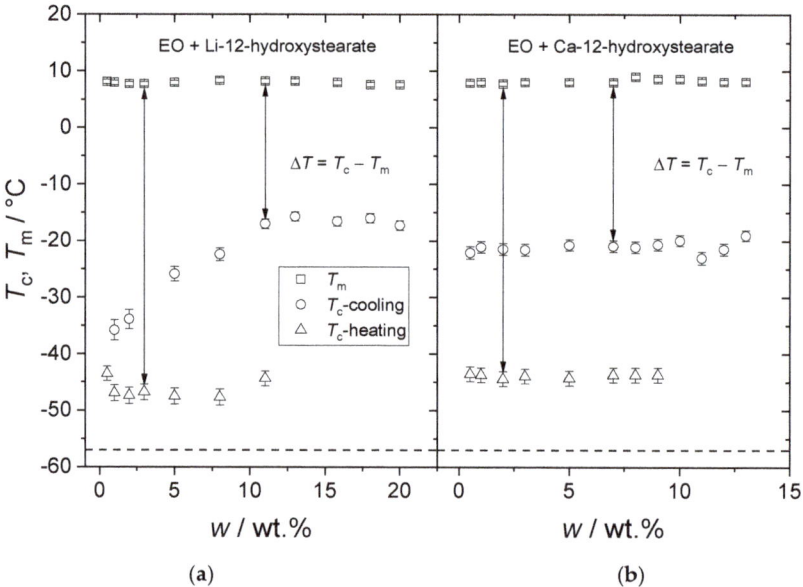

Figure 6. Crystallization temperatures obtained during cooling ($T_{c,\text{cooling}}$) and heating cycles ($T_{c,\text{heating}}$) and melting temperature T_m, as a function of thickener concentration w for Li-12-hydroxystearate (**a**) and Ca-12-hydroxystearate greases (**b**) based on trimelltiate (EO), obtained from the onset temperature of the exothermic peaks of DSC measurements with 2 K min^{-1} cooling and heating rate (Figure 5). The crystallization temperature at heating $T_{c,\text{heating}}$ is independent of the thickener fraction at about −45 °C, and the melting temperature is about 7.8 ± 0.4 °C. The dashed line marks the pour point of −57 °C.

The pure ester (EO) does not crystallize upon cooling but only upon heating ($T_{c,\text{heating}} = -43 \pm 0.3$ °C). The presence of thickener causes the oil to crystallize during both, cooling and heating cycles. For Li-12-hydroxystearate contents between 1 and 11 wt.% crystallization can be observed upon cooling and heating, for thickener contents above 11 wt.% only during cooling. Up to a thickener concentration of 11 wt.% the crystallization temperature $T_{c,\text{cooling}}$ increases linearly from −34 °C to −17 °C, but the crystallization temperature during heating $T_{c,\text{heating}}$ remains constant at −46.1 ± 1.9 °C. Above 11 wt.% the Li-12-hydroxystearate greases crystallize only during cooling and the crystallization temperature is $T_{c,\text{cooling}} = -16.7 \pm 0.7$ °C.

The trimellitate (EO) in Ca-12-hydroxystearate greases crystallizes at thickener concentrations between 0.5 and 9 wt.% during cooling and heating and above 9 wt.% only during cooling. With increasing thickener concentration, the crystallization temperature upon cooling increases slightly from -22.1 ± 1.1 °C to -19.0 ± 1.0 °C, whereas the crystallization temperature upon reheating remains constant at -43.7 ± 0.4 °C, irrespective of thickener concentration.

The enthalpy of fusion $\Delta h_f = 92.5 \pm 3.0$ J g^{-1} (endothermic, Figure 5b) is larger than the crystallization enthalpy $\Delta h_c = 73.5 \pm 6.0$ J g^{-1} because crystal growth is to some extent too slow to cause a detectable heat flux signal. Furthermore, the total enthalpy of fusion Δh_f and crystallization Δh_c are independent of the thickener concentration, indicating that only the base oil crystallizes.

Figure 7 shows the specific enthalpies of crystallization of the greases based on trimellitate (EO) during cooling $\Delta h_{c,cooling}$ (a) and heating $\Delta h_{c,heating}$ (b) relative to the total enthalpy of crystallization $\Delta h_c = \Delta h_{c,cooling} + \Delta h_{c,heating}$ as a function of Li- and Ca-12-hydroxystearate concentration.

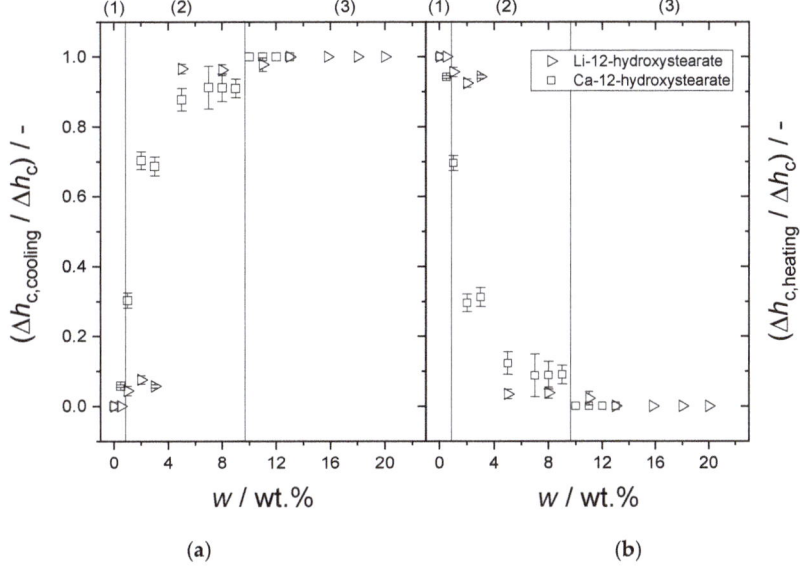

Figure 7. Specific enthalpies of crystallization during cooling $\Delta h_{c,cooling}$ (a) and heating $\Delta h_{c,heating}$ (b) normalized to the total enthalpy of crystallization $\Delta h_c = \Delta h_{c,cooling} + \Delta h_{c,heating}$ of the trimellitate (EO) based greases vs. concentration of Li- and Ca-12-hydroxystearate as determined from DSC measurements (see Figure 5) at cooling/heating rates of 2 K min^{-1}. The total enthalpy of crystallization remains constant at $\Delta h_c = 73.5 \pm 6.0$ J g^{-1} for both types of thickeners. Ranges (1–3) highlight the shift in the supercooling effect. Range (1) indicates predominantly cold crystallization, (2) crystallization upon cooling and heating, and (3) predominantly crystallization upon cooling.

According to ISO 11357-7 [20] the exothermic peaks' total area during cooling and heating corresponds to complete crystallization, and thus the percentage of crystallization during cooling and heating can be calculated. Figure 7 depicts that the percentage of crystallizing base oil upon cooling increases with increasing thickener concentration. As already shown in Figure 5, this indicates a monotonically increasing proportion of heterogeneous crystallization. At thickener concentrations >9 wt.% Ca-12-hydroxy stearate and >11 wt.% Li-12-hydroxystearate, respectively, only crystallization during cooling occurs because heterogeneous nucleation is predominant.

In the case of the pure trimellitate (EO), nuclei form below the temperature range in which crystal growth is optimal. In case of greases, the thickener creates interfaces at which nuclei can form in the temperature range in which crystal growth takes place. Since nuclei are present, crystal growth is detectable as an exothermic DSC signal [9–11].

At low thickener concentrations (0–9 wt.% for Ca-12-hydroxystearate and 0–11 wt.% Li-12-hydroxystearate) only a few nuclei are present. Therefore, only a few crystals start to grow, and the crystal growth rate is not fast enough for complete crystallization before reaching the lowest measuring temperature (see Figure 5a).

At thickener concentrations >9 wt.% for Ca-12-hydroxystearat and >11 wt.% for Li-12-hydroxystearate, sufficient amounts of nuclei are present to allow for complete crystallization of the ester upon cooling.

Due to heterogeneous crystallization, the lubrication greases crystallize approximately 37 K above the pour point. Thus, the pour point is not suitable to judge the low temperature application behavior of greases based on trimellitate (EO) with linear alkyl chains (C8–C10).

4. Conclusions

This study investigates the influence of thickener type and concentration on the crystallization, melting and glass transition of greases made from three different base oil types (Groups I–III) using rheological and thermoanalytical measurements. Group I is represented by a mineral oil (MO), Group II includes amorphously solidifying synthetic lubricating oils (PAG, KR-008, PAO8), and Group III comprises the crystallizing synthetic lubricating ester oil (EO).

4.1. Lubricating Greases Based on Mineral Oil (Group I)

During cooling below the crystallization temperature $T_{c,rheo}$, mineral oils precipitate paraffin crystals. Upon further cooling the complex viscosity reaches an absolute value of about 10^5 Pas, regardless of the thickener type and concentration. The suspension formed by the deposition of paraffin crystals is responsible for the increasingly stiff consistency of the lubricating greases. On reheating, above the melting temperature $T_{m,rheo}$, the absolute value of the complex viscosity returns to the initial level. The different thickener types (Li- or Ca-12-hydroxystearate) have a significant effect on crystallization (+4 K) and melting temperature (−4 K). The pour point (−12 °C) of the mineral oil is 3 K lower than the crystallization temperature of the lubricating grease and 7 K lower than the melting temperature. This indicates that the pour point does not properly define the lowest application temperature for mineral oil (MO) based greases.

4.2. Lubricating Greases Based on Non-Crystallizing Synthetic Oils (Group II)

Group II oils do not crystallize but solidify glass-like at temperatures below −70 °C. This group includes polyalphaolefin (PAO8), alkylated naphthalene (KR-008), and polypropylene glycol (PAG). The addition of Li-12-hydroxystearate does not change the glass transition temperature significantly. The absolute value of the complex viscosity of these greases increases steadily with decreasing temperature. At about 20 K above T_G, however, the absolute value of the complex viscosity increases sharply, i.e., the slope of the $|\eta^*|$ vs. T curve changes significantly. The onset of this steep viscosity increase is equal to the pour point of the corresponding oil and the latter thus is an appropriate measure for the lowest application temperature of these greases.

4.3. Lubricating Greases Based on Crystallizing Synthetic Oils (Group III)

Group (III) oils crystallize with strong supercooling effects up to cold crystallization. Trimellitate (EO) with linear C8-C10 alkyl chains belongs to this group. This ester oil (EO) shows that adding minor thickener concentrations (about 0.5 wt.% Ca-12-hydroxystearate, 1 wt.% Li-12-hydroxystearate) causes the ester to crystallize partly upon cooling and partly upon heating. The fraction of crystallization during cooling rises with increasing thickener concentration until the ester oil crystallizes only upon cooling. This is due to heterogeneous nucleation, which becomes more significant with increasing thickener concentration. As a result of heterogeneous nucleation, the crystallization temperature rises from $-45\,°C$ to $-20\,°C$. In contrast, the melting temperature T_m remains constant at $7\,°C$ independent of the thickener concentration and thickener type. The investigations show that using lubricating greases based on this ester oil below the melting temperature is not advisable. Especially for lubricating greases based on trimellitate (EO), the pour point of $-57\,°C$ is inappropriate as indicator for the lowest application temperature.

Author Contributions: Conceptualization, A.C. and K.-H.J.; methodology, A.C., A.H., B.H., K.-H.J. and N.W.; validation, A.C., A.H. and B.H.; formal analysis, A.C. and A.H.; investigation, A.C. and A.H.; resources, K.-H.J. and N.W.; data curation, A.C.; writing—original draft preparation, A.C.; writing—review and editing, A.C., A.H., B.H., K.-H.J. and N.W.; visualization, A.C.; supervision, K.-H.J. and N.W.; project administration, B.H., K.-H.J. and N.W.; funding acquisition, K.-H.J. and N.W. All authors have read and agreed to the published version of the manuscript.

Funding: This research was funded by Arbeitsgemeinschaft industrieller Forschungsvereinigungen „Otto von Guericke" e.V. (AiF), Grant Number 20001N and Forschungsvereinigung Antriebstechnik e.V. (FVA), Grant Number 829I.

Acknowledgments: The authors give thanks to the members of the FVA, especially to Fuchs Schmierstoffe GmbH, Castrol Germany GmbH, and King Industries, for providing additional information and the base oils. In addition, we wish to thank the students Johanna Scheller, Daniel Weiß, and Maximilian Enhuber for their contribution to the manuscript as part of their bachelor thesis.

Conflicts of Interest: The authors declare no conflict of interest.

Appendix A

Table A1. Base oil, thickener type, mass concentrations of thickener, fulling penetration P_W [17], and corresponding NLGI class [18] for the lubricating greases.

Base Oil	Thickener	$w_{thickener}$/wt%	P_W/0.1 mm	NLGI Class
MO	Li-12-hydroxystearate	2.0	n.a	n.a
		5.0	430 ± 6	000/00
		10.0	362 ± 9	0
		15.0	264 ± 2	2/3
		20.0	212 ± 2	3/4
MO	Ca-12-hydroxystearate	0.5	n.a.	n.a.
		1.0	n.a.	n.a.
		2.0	n.a.	n.a.
		3.0	n.a.	n.a.
		5.0	n.a.	n.a.
		7.8	454 ± 6	000
		10.0	427 ± 6	00
		11.0	335 ± 3	1
		12.1	244 ± 3	3
		13.0	221 ± 2	3
		14.0	204 ± 5	4
		15.8	195 ± 2	4
PAO8	Li-12-hydroxystearate	22.0	286 ± 4	2
KR-008		10.0	281 ± 7	2
KR-008		11.0	247 ± 1	3
PAG		11.0	306 ± 5	1/2
PAG		15.0	271 ± 12	2
PAO8	Ca-12-hydroxystearate	22.0	227 ± 9	3
KR-008		11.0	244 ± 1	3
PAG		11.0	231 ± 5	3

Table A1. Cont.

Base Oil	Thickener	$w_{thickener}$/wt%	P_W/0.1 mm	NLGI Class
EO	Li-12-hydroxystearate	0.5	n.a.	n.a.
		1.0	n.a.	n.a.
		2.0	n.a.	n.a.
		3.0	n.a.	n.a.
		5.0	n.a.	n.a.
		8.0	n.a.	n.a.
		11.0	442 ± 6	000/00
		13.0	361 ± 5	0
		15.8	272 ± 5	2
		18.0	252 ± 1	2/3
		20.0	183 ± 10	4
EO	Ca-12-hydroxystearate	0.5	n.a.	n.a.
		1.0	n.a.	n.a.
		2.0	n.a.	n.a.
		3.0	n.a.	n.a.
		5.0	n.a.	n.a.
		7.0	438 ± 12	000/00
		8.0	437 ± 8	000/00
		9.0	434 ± 8	000/00
		10.0	316 ± 3	1
		11.0	254 ± 2	2/3
		12.0	242 ± 4	3
		13.0	253 ± 5	2/3

n.a.: not applicable

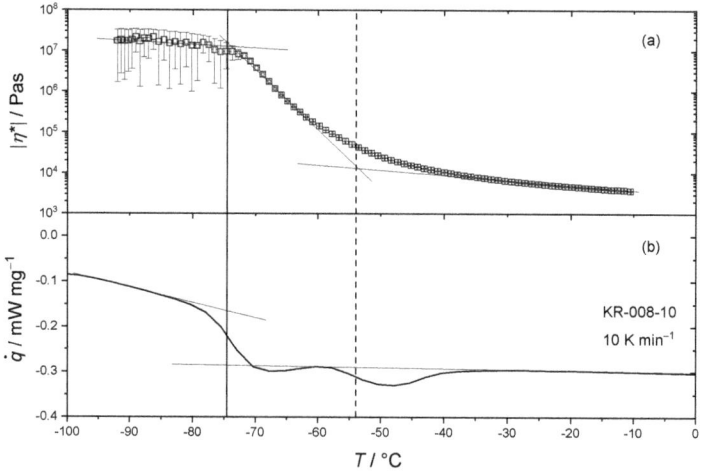

Figure A1. (a) Complex viscosity $|\eta^*|$ as a function of the temperature of the lubricating grease made from the base oil KR-008 with 10 wt.% Li-12-hydroxystearate as a thickener (KR-008-10), measured in oscillatory shear with the deformation $\gamma = 0.05\%$, angular frequency $\omega = 10$ rad s^{-1} and cooling rate of -10 K min^{-1}. Lines are to guide the eye. (b) Specific heat flow $\dot{q}$ of the KR-008-10 grease at a heating rate of 10 K min^{-1}. The grease shows a glass transition temperature T_G at -74.5 ± 0.4 °C, marked by a solid line. The dashed line marks the pour point for the base oil KR-008 of -54 °C.

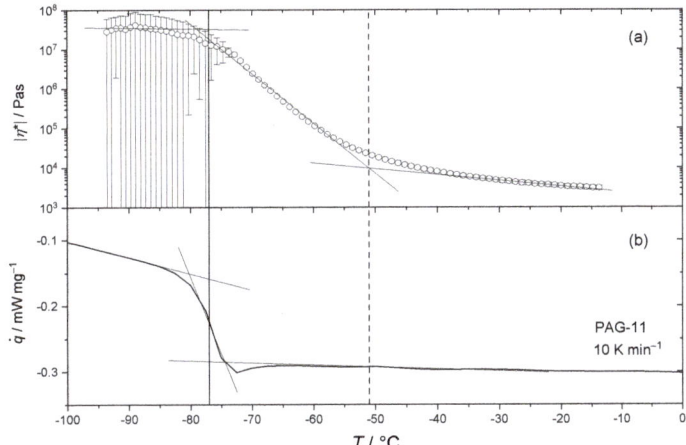

Figure A2. (**a**) Complex viscosity $|\eta^*|$ as a function of temperature of the lubricating grease made from the base oil PAG with 11 wt.% Li-12-hydroxystearate as a thickener (PAG-11), measured in oscillatory shear with the deformation $\gamma = 0.05\%$, angular frequency $\omega = 10$ rad s^{-1} and cooling rate of -10 K min^{-1}. Lines are to guide the eye. (**b**) Specific heat flow $\dot{q}$ of the PAG-10 grease at a heating rate of 10 K min^{-1}. The grease shows a glass transition temperature T_G at -77 ± 0.2 °C, marked by a solid line. The dashed line marks the pour point for the base oil PAG of -51 °C.

References

1. De Laurentis, N.; Cann, P.; Lugt, P.M.; Kadiric, A. The Influence of Base Oil Properties on the Friction Behaviour of Lithium Greases in Rolling/Sliding Concentrated Contacts. *Tribol. Lett.* **2017**, *65*, 128. [CrossRef]
2. Lan, Z.; Liu, S.; Xiao, H.; Wang, D. Frictional Behavior of Wax–Oil Gels Against Steel. *Tribol. Lett.* **2017**, *65*, 88. [CrossRef]
3. Conrad, A.; Hodapp, A.; Hochstein, B.; Willenbacher, N.; Jacob, K.-H. Low-Temperature Rheology and Thermoanalytical Investigation of Lubricating Oils: Comparison of Phase Transition, Viscosity, and Pour Point. *Lubricants* **2021**, *9*, 99. [CrossRef]
4. D02 Committee. *ASTM D4693: Test Method for Low-Temperature Torque of Grease-Lubricated Wheel Bearings*; ASTM International: West Conshohocken, PA, USA, 2017.
5. D02 Committee. *ASTM D1478: Test Method for Low-Temperature Torque of Ball Bearing Grease*; ASTM International: West Conshohocken, PA, USA, 2020.
6. D02 Committee. *ASTM D97 Test Method for Pour Point of Petroleum Products*; ASTM International: West Conshohocken, PA, USA, 2017.
7. Webber, R.M. Low-Temperature rheology of lubricating mineral oils: Effects of cooling rate and wax crystallization on flow properties of base oils. *J. Rheol.* **1999**, *43*, 911–923. [CrossRef]
8. Li, S.; Bouzidi, L.; Narine, S.S. Lubricating and Waxy Esters, V: Synthesis, Crystallization, and Melt and Flow Behaviors of Branched Monoesters Incorporating 9-Decenol and 9-Decenoic Acid. *Ind. Eng. Chem. Res.* **2014**, *53*, 12339–12354. [CrossRef]
9. Turnbull, D. Kinetics of Heterogeneous Nucleation. *J. Chem. Phys.* **1950**, *18*, 198–203. [CrossRef]
10. Glicksman, M.E. *Principles of Solidification: An Introduction to Modern Casting and Crystal Growth Concepts*; Springer Science & Business Media LLC: New York, NY, USA, 2011; ISBN 9781441973436.
11. March, N.H.; Street, R.A.; Tosi, M.P. *Amorphous Solids and the Liquid State*; Springer: Boston, MA, USA, 1985.
12. Delgado, M.A.; Sánchez, M.C.; Valencia, C.; Franco, J.M.; Gallegos, C. Relationship Among Microstructure, Rheology and Processing of a Lithium Lubricating Grease. *Chem. Eng. Res. Des.* **2005**, *83*, 1085–1092. [CrossRef]
13. Delgado, M.A.; Valencia, C.; Sánchez, M.C.; Franco, J.M.; Gallegos, C. Influence of Soap Concentration and Oil Viscosity on the Rheology and Microstructure of Lubricating Greases. *Ind. Eng. Chem. Res.* **2006**, *45*, 1902–1910. [CrossRef]
14. Delgado, M.A.; Secouard, S.; Valencia, C.; Franco, J.M. On the Steady-State Flow and Yielding Behaviour of Lubricating Greases. *Fluids* **2019**, *4*, 6. [CrossRef]
15. Yeong, S.K.; Luckham, P.F.; Tadros, T.F. Steady flow and viscoelastic properties of lubricating grease containing various thickener concentrations. *J. Colloid Interface Sci.* **2004**, *274*, 285–293. [CrossRef] [PubMed]
16. D02 Committee. *ASTMD7346.9011: Standard Test Method for No Flow Point and Pour Point of Petroleum Products and Liquid Fuels*; ASTM International: West Conshohocken, PA, USA, 2015.
17. DIN Deutsches Institut für Normung e.V. *DIN ISO 2137: Petroleum Products and Lubricants—Determination of Cone Penetration of Lubricating Greases and Petrolatum*; Beuth Verlag GmbH: Berlin, Germany, 2007.

18. DIN Deutsches Institut für Normung e.V. *DIN 51818 Lubricants;* Consistency Classification of Lubricating Greases; Beuth Verlag GmbH: Berlin, Germany, 1981.
19. Mortimer, R.G. *Physical Chemistry*, 2nd ed.; Harcourt/Academic Press: San Diego, CA, USA, 2000; ISBN 9780080538938.
20. DIN Deutsches Institut für Normung e.V. *Differential Scanning Calorimetry (DSC)—Part 7: Determination of Crystallization Kinetics (ISO 11357-7:2015);* Beuth Verlag GmbH: Berlin, Germany, 2015.

Article

Tribological Behavior of Novel CNTs-Based Lubricant Grease in Steady-State and Fretting Sliding Conditions

Adolfo Senatore [1,*], Haiping Hong [2], Veronica D'Urso [1] and Hammad Younes [2]

[1] Department of Industrial Engineering, University of Salerno, via Giovanni Paolo II, 132, I-84084 Fisciano, Italy; v.durso3@studenti.unisa.it
[2] Department of Electrical Engineering, South Dakota School of Mines and Technology, Rapid City, SD 57701, USA; Haiping.Hong@sdsmt.edu (H.H.); Hammad.Younes@sdsmt.edu (H.Y.)
* Correspondence: a.senatore@unisa.it

Abstract: The tribological behavior of novel 7.5 wt% carbon nanotube-based lubricant greases in PAO (polyalphaolefin) oil with and without 1.0 wt% MoS_2, together with several other commercial greases such as calcium, lithium, were studied. The test results showed a marked reduction of frictional coefficient achieved by the CNTs based grease samples with an average benefit of around 30% compared to conventional greases. The steady state test under 1.00 GPa average contact pressure in a mixed lubrication regime and the fretting test showed the best results in terms of friction reduction obtained by CNTs greases. Steady state tests at higher average contact pressure of 1.67 GPa proved to have a lower friction coefficient for CNTs grease containing MoS_2; otherwise CNTs grease without MoS_2 showed an average value of CoF comparable to calcium and lithium greases, both in a boundary and a mixed regime. The protection against wear, a considerable decrease (−60%) of reference parameter was measured with CNTs grease with MoS_2 (NLGI 2) in comparison with the worst conventional grease and −22% in comparison with the best conventional grease. The data indicated that our novel carbon nanotube greases show superior tribological properties and will have promising applications in the corresponding industry.

Keywords: grease; PAO; CNTs; wear; coefficient of friction; tribological behavior; fretting; sliding

1. Introduction

Most of the failures and energy losses of mechanical systems are due to friction and wear, so lubrication could be considered one of the most effective ways to reduce surface damage and energy dissipations. Lubricants decrease wear and friction in mutual contacts between coupled surfaces. Several types of additives in lubricants (e.g., Extreme pressure (EP), antiwear (AW), Frequency-modulation (FM)) yield specific requirements for good lubrication. The previous research proved that lubricants' tribological properties have been improved using nanoparticles as a new additive [1]. Nanoadditives enhance the lubricating characteristics due to the tiny particles' size and morphology that can fill the surface asperities and realize the lubrication mechanisms (i.e., rolling effect, mending effect, polishing effect, protective film formation [1,2]). Research has focused on several typologies of nanomaterials, such as chalcogenides, metal oxides, and carbon-based additives in oil or grease [3–9]. For instance, experimental research has proved that metal oxides added to lithium grease can significantly improve the lubricants' tribological behavior. As reported in [10], the lubricating properties of grease specimens with different concentrations of Al_2O_3 nanoparticles have been investigated using a pin-on disc apparatus. Results have shown a reduction of COF and wear scar width by approximately 57.9% and 47.5%, respectively. Moreover, research has been carried out on the effects of agglomeration of selected nanoparticles, such as zirconia, within a lithium grease [11]. Experimental analysis on friction properties has proved the addition of 1 wt% ZrO_2 nanoparticles to pure lithium grease can decrease the friction coefficient to 50%. Nevertheless, the agglomeration of

ZrO$_2$ nanoparticles in the lithium grease can increase the friction coefficient by two times compared to that for the pure grease. Applying nanomaterials and nanotechnology in lubrication has also become increasingly popular to achieve green manufacturing and its sustainable development. To this end, water-based nanolubricants are emerging as promising alternatives to the traditional oil-based lubricants [12].

As nanomaterials are of growing interest for lubricating systems, more and more attention has been given to carbon nanotubes (CNTs). Their tubular structure exhibits excellent mechanical characteristics (i.e., high tensile strength, high elastic modulus [13,14], high thermal and electrical conductivity [15–18]) and good lubrication properties, enhancing wear and coefficient Friction (CoF) reduction [19].

In the last few years, the use of carbon nanotubes as additives in lubricating oil/grease has increased, and many research studies have proved that CNTs, with the single-wall (SWCNT) or multi-wall (MWCNT) structure, can improve the tribological performances of the lubricant significantly, compared to a traditional one. For instance, the tribological behavior of Co-based single wall carbon nanotubes has been investigated as an antiwear and extreme pressure additive to SAE base oil. Friction and wear have been evaluated, and it has been found that the SWCNTs are more efficient than the commercial oil additives; tests proved the decreasing of the friction coefficient, using only 0.5 wt% carbon nanotubes to the base oil [20]. Different concentrations of multi-wall carbon nanotubes have been added as additives to two mineral oils; the samples were tested for their antiwear, load carrying capacity, and friction coefficients. The experimental results have shown that the addition of MWCNTs to base oils exhibited good friction reduction and antiwear properties: wear and friction have decreased by −68%/−39% and −57%/−49%, respectively, in the case of MWCNTs-based mineral oils compared with the two reference mineral oils [21].

A weighted percentage of 1 wt% of carbon nanotubes in a traditional lithium-based grease can improve EP properties, load-carrying capacity, AW, and friction performances [22]. Moreover, the CNTs concentration affects the lubricant's tribological characteristics, as shown in [23], whereas the different percentage of nanoparticles has been added to a common calcium-based grease. Carbon nanotubes have been added together with other nanoparticles in a calcium grease, and improvements have been observed not only in reduced wear and friction but also in other important characteristics of a lubricant grease, such as shear stress, apparent viscosity, drop point, and thermal conductivity [24–26].

CNTs have been used with different nanoparticles and materials to improve the tribological behavior of lubricants [27–29]. Wang et al. coated the surface of carbon nanotubes with a uniform copper nanoparticle. This successful coating for the nanocomposite was achieved through the modification of the CNTs surface with spontaneous polydopamine (PDA). The friction and wear were reduced by 33.5% and 23.7% when 0.2 wt% of the prepared nanocomposite was used. The improvement in the lubrication properties was attributed to uniform coating and the formation of transfer films on the rubbing surfaces [30]. Akbarpour et al. studied the wear and friction of a composite made of MWNTs, copper, and SiC. The result revealed that 2 vol% of SiC and 4 vol% CNT and Copper had the highest wear resistance and a lower friction coefficient [31]. Song et al. used the chemical vapor deposition technique to grow several layers of MoS$_2$ on the surface of the carbon nanotubes. The CNTs were oxidized by acids to prepare strongly oxidized CNTs (SOCNTs). 0.02 wt% of SOCNTs with dibutyl phthalate (DBP) achieved a reduction in the friction coefficient and wear scar diameter by 57.93% and 19.08%, respectively [32].

According to the ASTM D288, the definition of lubricating grease is "a solid to a semi-fluid product of a thickening agent in a liquid lubricant. Other ingredients imparting special properties may be included" [33]. The thickener agent gives a solid to semi-fluid structure to the lubricant grease and affects many of its main properties. Recently, a new class of nanogrease has been defined using carbon nanotubes as a thickener phase in the lubricant. SWCNTs and MWCNTs have been added into the polyalphaolefin (PAO) oils to form stable and homogeneous CNT grease with potential heat transfer, conductive, and lubricative properties. The rheological investigation of these CNT greases has provided information concerning the formation of a stable tridimensional structure and particle-particle interactions of CNT suspensions, explaining its excellent lubricating performances [34]. The characterizations of four selected CNT greases have shown high dropping temperatures, very low oil leaking percentages even at high temperatures, and a substantial increase in thermal and electrical conductivities [35]. Experimental tests have been carried out to investigate the enhancing tribological properties of CNT grease compared to the base oil grease. It has been showing that CNTs play a more significant part in lubrication, which greatly improves the lubricating effect on wear and friction [36–39].

In this research, CNTs, single-wall and multi-wall (SWCNTs and MWCNTs) have been added separately into polyalphaolefin oils (DURASYN_166), with and without MoS_2, to form stable and homogeneous CNT greases. The tribological behavior has been investigated in steady-state and fretting sliding conditions, and performance comparisons have been conducted with commercial calcium and lithium greases.

2. Experimental Section

2.1. Materials

SWCNTs and MWCNTs were purchased from Carbon Nanotechnologies Incorporation (CNI, Houston, TX, USA now Unidym Inc). Carbon nanotubes were used as received with no further purification or functionalization. SWCNTs are of average diameter 1.4 nm and are found in "ropes" which are typically *20 nm in diameter or *50 tubes per rope with lengths of 0.5–40 microns. Thermal conductivity is around 35 W/mK at room temperature. No electrical conductivity data are available. The sample purity is 70–90 vol%. There are 10–25 vol% carbon black and <5 vol% residual catalyst metal impurities in the samples. For MWCNTs, the diameter is around 20–40 nm, and the standard length is around 0.5–40 microns. No thermal and electrical conductivity data are available. The sample purity is around 90 vol%. There are <10 vol% carbon black (amorphous carbon) and a small amount (<0.5 vol%) of catalyst metal impurities in the samples. DURASYN 166 is a commercial polyalphaolefin oil product purchased from Chemcentral (Chicago, IL, USA). MoS_2 nanoparticles were purchased from Sigma Aldrich (St. Louis, MO, USA). Since traditional calcium and lithium greases are commonly employed in industrial lubrication, they were chosen in order to set a comparison to the tribological performance of CNTs grease. Commercial samples of Eurogrease I.G. NLGI 2 calcium based, Eurogrease C.M.3 NLGI 3 lithium based and Eurogrease G.O. NLGI 2 lithium based with MoS2 additives from Rilub SpA (Ottaviano, Italy) were used as reference.

2.2. Grease Fabrication

Sonication was performed using a Branson Digital Sonifier, model 450. A three-roll mill by Ross Engineering Inc. (Savannah, GA, USA) was used to incorporate the CNTs to make the stable and well-dispersed greases. The carbon nanotube greases were made by adding a calculated amount of nanotubes (single-wall or multi-wall) into a beaker that contained an appropriate amount of oil (e.g., DURASYN 166, Polyalphaolefin). 5 wt% MoS_2 was added to carbon nanotube grease to see how the additive would influence the physical property. In this paper, we use two greases: singe-wall carbon nanotube (SWCNTs) grease and multi-wall carbon nanotube (MWCNTs) grease with MoS_2. The experimental procedure to make nano grease is as follows: first, the fluids were sonicated 5–10 times, each time 1–2 min. Second, the fluids were poured into a three-roll mill, repeated 5–8 times. Finally, black, stable, and well-dispersed greases were manufactured and collected.

2.3. Tribological Characterization

The tribological test rig for exploring nano additives' effectiveness is the Wazau TRM 100 (Dr.-Ing. Georg Wazau Meß Prüfsysteme GmbH, Berlin, Germany) with the ball-on-rotational disc setup as in the drawing in Figure 1. The selected mating materials are: (ball) steel X45Cr13, hardness: HRC 52-54, diameter: 8 mm, mean surface roughness:

0.2 µm; (disc) steel X155CrVMo12-1, hardness: HRC 60, diameter: 105 mm, mean surface roughness: 0.5 µm.

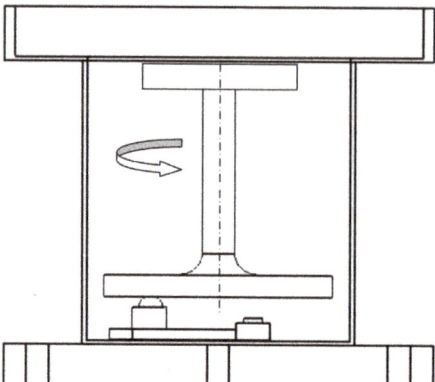

Figure 1. Ball on disk setup and rotational tribometer.

The tribological tests were carried out to investigate two different lubrication regimes and fretting operating conditions. A normal load level of 20 or 95 N was applied to attain an average hertzian contact pressure around 1.00 and 1.67 GPa, respectively. The average grease temperature was kept constant at room temperature in each test. The relative motion between the steel ball and the disc was pure sliding at two-speed levels: 5 and 500 mm/s. According to previous results for the same tribopair geometry, average contact pressure, temperature, the boundary lubrication regimes, and the mixed lubrication superposition of the boundary, and elastohydrodynamic lubrication (EHL) regimes were covered. The fretting test was designed by applying an alternate rotating motion to the disk to cover a 75° angle with 2 Hz as the frequency.

Both the ball and the disk were pre-cleaned, and the grease was then evenly pasted on the disk sliding pathway by forming a layer with 2 mm thickness. The test length was 60 min in steady-state condition and 120 min for fretting. Thus, the actual sliding distance was 18 and 1800 m for the speed of 5 and 500 mm/s, respectively. The sliding distance was 1131 m for the fretting portion/phase of testing.

The friction coefficient has been indirectly measured in real-time using a torque sensor located under the ball holder plate. After each tribotest, the wear damage circle on the top of the worn steel ball was offline measured through a 3D confocal microscopy.

3. Results and Discussion

3.1. Tribological Results

The results of the frictional tests were performed under the operating conditions described above in Section 2.3. In both, the investigated lubrication regimes are summarized in Table 1. Tests were also performed to analyze the behavior of the lubricating grease samples for the wear damage of the steel ball surface. In particular, the worn surface of the steel ball was analyzed using a 3D confocal microscope to measure the wear scar diameter (WSD). According to the ISO/IEC Guide 98-3:2008, the expanded uncertainty of the frictional data and WSD measurements are 5.0×10^{-3} and 1 µm, respectively, by assuming the coverage factor k = 2.

The high frequency acquired frictional data were processed through a rolling mean to filter out spike values and typical oscillations of such measurements.

The steady-state test at 5 mm/s of sliding speed and initial average hertzian pressure equal to 1.00 GPa provided the graphs collected in Figure 2. The graphs show CoF values in a narrow range of 0.12–0.14, with high similarities between Calcium soap, Lithium soap, and CNT-based grease (0.13 as average). The two remaining samples, i.e., Lithium soap with MoS$_2$ and CNTs with MoS$_2$ provided lightly higher CoF. This outcome confirms the effectiveness of MoS$_2$ as an additive in oil and lubricant grease in extreme–pressure conditions, tougher than non-critical conditions as from this test.

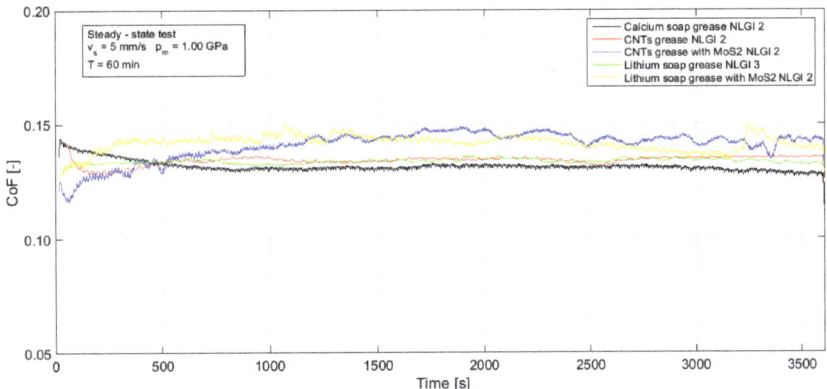

Figure 2. Friction coefficient in boundary regime under 1.00 GPa average contact pressure.

In Figure 3, the increasing value of sliding speed (500 mm/s) results in a marked difference in the steady-state value of the friction coefficient. Except for the Calcium soap grease, showing high CoF variability in the first 25 min of the test, all the samples reach their asymptotic value quickly. The former result is not unexpected as the Calcium soap grease with NLGI grade 2 does not bear sliding condition with speed as high as 500 mm/s; the latter behavior underlines a good reduction achieved by the CNTs based samples with CoF in the range 0.11–0.12 with a reduction up to −35% compared to conventional greases over the second part of the test characterized by stationary behavior.

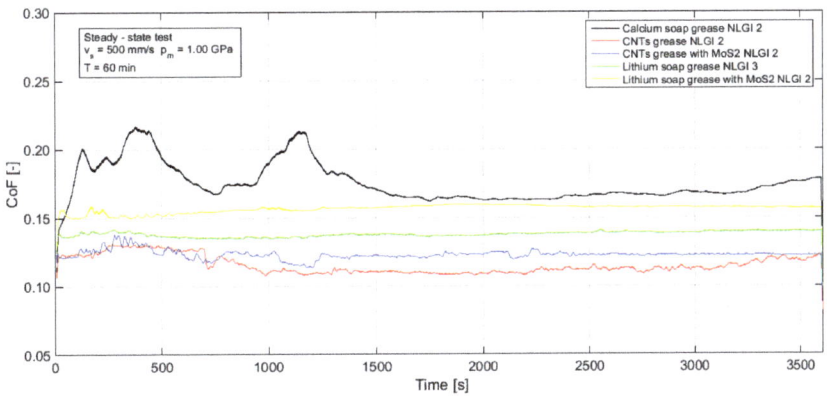

Figure 3. Friction coefficient in mixed regime under 1.00 GPa average contact pressure.

The measurements at a sliding speed of 5 mm/s and initial average hertzian pressure of 1.67 GPa are shown in Figure 4. In these conditions, the friction coefficient values are in the range 0.10–0.13, and for this test, the trend of the graphs is relatively constant. Again, the CoF values are comparable for some samples, such as Lithium soap, Lithium soap with MoS$_2$ additive, and CNTs based grease. Interesting performances are observable for the Calcium soap grease and CNTs based grease with MoS$_2$ additive, whose CoFs are around 0.10. This behavior is expected for these tested samples. As previously explained, the

Calcium based grease allows good tribological behavior in non-critical working conditions, e.g., at low temperatures and low speeds, whereas the properties of molybdenum bisulphide and carbon nanotubes, combined in the grease sample increase its tribological characteristics.

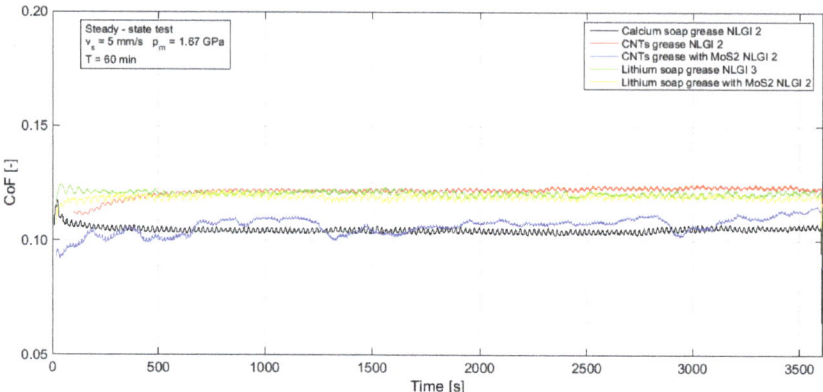

Figure 4. Friction coefficient in boundary regime under 1.67 GPa average contact pressure.

Furthermore, all samples exhibited decreased friction coefficient values while increasing normal load, demonstrating good lubricating capacity under high loads and low speeds in a boundary lubrication regime. As a matter of fact, the friction coefficient showed a tendency to decrease with increasing contact pressure. According to previous works [3,4], the shear stress increased less rapidly in proportion to the contact pressure. This leads to a reduction of the friction coefficient with increasing pressure at a given level of sliding speed.

Among the frictional tests discussed in this paper in a steady-state condition, the most demanding combination of sliding speed and load is given by the test in a mixed lubrication regime obtained at 500 mm/s and 1.67 GPa, seen in Figure 5. Once again, it is worth noting that the trend of the graphs in Figure 5 is nearly constant for all samples, except for Calcium soap grease with CoF peaks as high as 0.27, as is further confirmed in the discussion regarding the test at low speed and load, Figure 4.

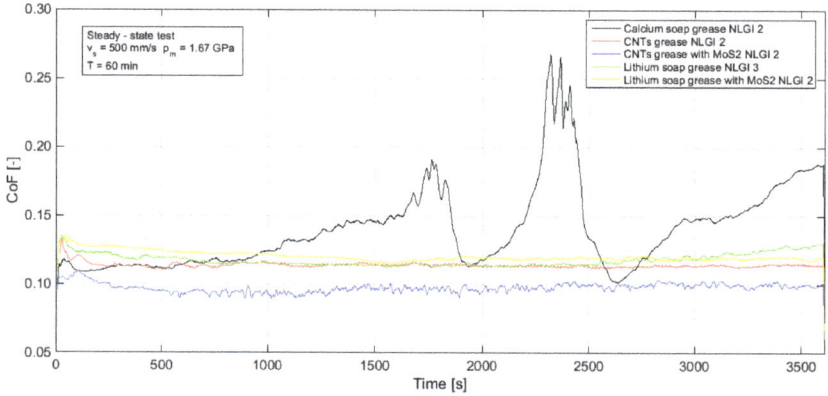

Figure 5. Friction coefficient in mixed regime under 1.67 GPa average contact pressure.

The samples of Lithium soap, Lithium soap with MoS$_2$, and CNTs based on the CoF curves are approximately superimposable, presenting CoF falling in the same range of the boundary test at 1.67 GPa, 0.12–0.13. In contrast, the CNTs based grease with MoS$_2$ showed a noticeable reduction in CoF: −28% compared to the three previous samples, by attaining an outstanding reduction to 0.09.

The last set of graphs collected in Figure 6 includes measurements of frictional behavior exhibited by the five grease samples under the characteristic fretting condition by applying sliding alternate motion to the frictional conjunction with frequency 2 Hz. The test length was 120 min with a covered sliding distance of 1131 m. In fretting conditions, unlike the previous ones in which the sliding speed is constant, stability at high speed is not expected for each CoF graph since the oscillating conditions may require a longer running-in time for the tribopairs' steel surfaces as well as the grease thickener structure. Along with the already exhibited lack of stationarity of the Calcium soap grease, even the Lithium soap and Lithium soap with MoS$_2$ additive show high variability of CoF over the whole 2-h test with average values attaining 0.15, 0.13, respectively.

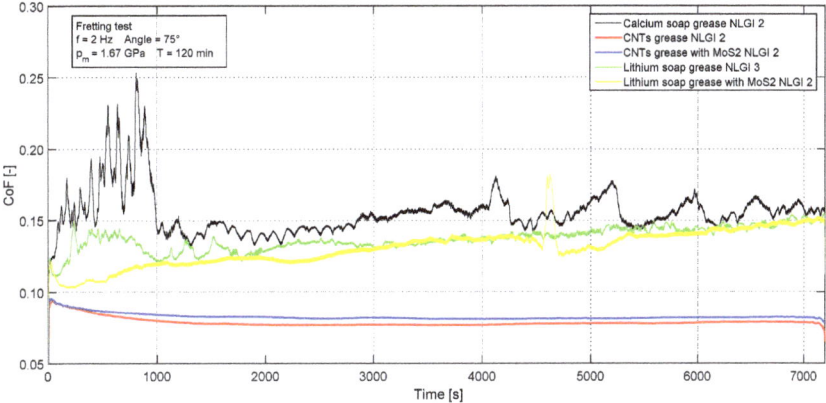

Figure 6. Friction coefficient in the fretting test under 1.67 GPa average contact pressure.

The CNTs based greases show lower CoF values, with an average value under 0.09 and a remarkable reduction of −35% with respect to the best performing conventional grease. Even the amplitude of CoF oscillations in testing CNTs based grease is drastically reduced by showing enhanced structural stability of the thickener and overall industrial reliability of the lubricant.

The frictional and wear data also correlate well, resulting in a positive correlation (0.8 Pearson index), Tables 1 and 2. This outcome is not obvious in such test conditions. Table 3 underlines the outstanding results achieved by using CNTs based grease with wear parameter reduction close to −60% in comparison with the worst conventional grease and −22% in comparison with the best conventional grease.

Table 1. The average friction coefficient in boundary and mixed regime.

Scheme	Friction Coefficient in Boundary Regime [-]		Friction Coefficient in Mixed Regime [-]	
	Contact Pressure 1.00 GPa	Contact Pressure 1.67 GPa	Contact Pressure 1.00 GPa	Contact Pressure 1.67 GPa
Ca soap grease NLGI 2	0.13	0.11	0.15	0.16
Li soap grease NLGI 3	0.13	0.13	0.14	0.12
Li soap grease with MoS_2 NLGI 2	0.14	0.12	0.15	0.13
CNTs grease NLGI 2	0.13	0.12	0.12	0.12
CNTs grease with MoS_2 NLGI 2	0.14	0.13	0.11	0.11

Table 2. Wear scar diameter (WSD) in boundary and mixed regime.

Sample	Wear Scar Diameter (WSD) in Boundary Regime [μm]		Wear Scar Diameter (WSD) in Mixed Regime [μm]	
	Contact Pressure 1.00 GPa	Contact Pressure 1.67 GPa	Contact Pressure 1.00 GPa	Contact Pressure 1.67 GPa
Ca soap grease NLGI 2	228	385	407	808
Li soap grease NLGI 3	630	268	358	557
Li soap grease with MoS_2 NLGI 2	321	307	398	551
CNTs grease NLGI 2	195	333	260	930
CNTs grease with MoS_2 NLGI 2	231	333	300	556

Table 3. The friction coefficient and wear scar diameter in the fretting test.

Sample	Friction Coefficient in Fretting Test [-]	Wear Scar Diameter (WSD) in Fretting Test [μm]
Ca soap grease NLGI 2	0.16	915
Li soap grease NLGI 3	0.15	547
Li soap grease with MoS_2 NLGI 2	0.13	481
CNTs grease NLGI 2	0.08	407
CNTs grease with MoS_2 NLGI 2	0.09	374

3.2. Surface Analysis Results

The SEM images were acquired on the worn surfaces of steel ball specimens after the fretting test. By comparing SEM-XRD spectra collected on the worn surface of the steel ball specimen after tribological tests with conventional greases and CNTs based grease, a significant difference in carbon content was found on the steel surface. As shown in Figure 7, the carbon content on the steel ball's worn surface after test with Calcium soap grease is around 15%, and other SEM measurements confirm this on a different portion of the worn surface.

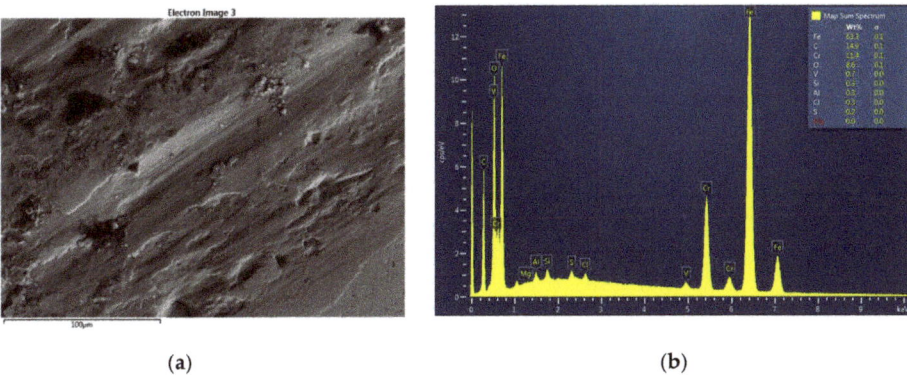

Figure 7. (a) SEM image of a worn surface ball of steel ball: Calcium soap grease; (b) EDX spectrum: Calcium soap grease.

The same analysis performed on the steel ball's worn surface post-tribological test with CNTs based grease shows lower carbon content (around 6–8%), Figure 8. In the same spectrum, it is worth noting the presence of Molybdenum.

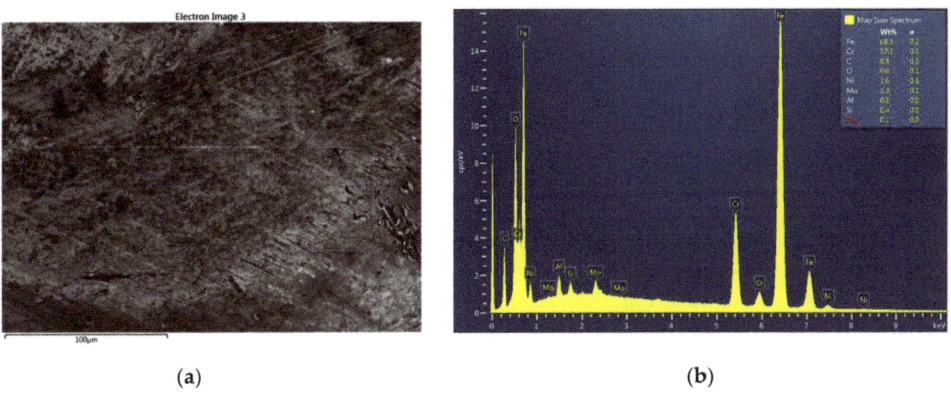

Figure 8. (a) SEM image of a worn surface ball of steel ball: CNTs based grease with MoS_2; (b) EDX spectrum: CNTs based grease with MoS_2.

The well-known literature analysis of the friction reduction mechanism introduced by nanoparticles as lubricant additives finds in the following list the more convincing physical explanations: rolling-sliding "rigid" motions together with flexibility properties, nano additive exfoliation, and material transfer to the metal surface to form the so called "tribofilm" or "tribolayer", electronic effects in tribological interfaces, surface roughness improvement effect or "mending"; along with the more classical hypothesis of surface sliding on lower shear stress layers due to weak interatomic forces, valid also for micro-scale additives used for decades. In the case of the samples proposed in this paper, along with the transfer of nanoparticles from semifluid lubricant to steel mating surfaces, CNTs used as sole thickener seem to provide to the bulk structure of the grease a marked stability of the tribological response over the whole spectrum of the performed tests from a qualitative point of view. From a quantitative point of view, frictional reduction is always good, the CoF has a marked reduction in fretting conditions. On the other hand, wear reduction is substantial only in the case of the fretting test and further optimization may be needed to increase wear protection in a broader range of load/speed combinations. The outstanding behavior in fretting conditions could be attributed to a more consistent and fatigue-resistant performance of CNTs thickener in comparison to conventional greases based on metallic soap.

4. Conclusions

The tribological behavior of novel carbon nanotube-based lubricant greases in PAO with and without MoS_2 was studied. The test results showed a marked reduction of frictional coefficient achieved by the CNTs based grease samples with an average benefit of around −30% compared to conventional greases. The steady state test under 1.00 GPa average contact pressure in a mixed lubrication regime and the fretting test showed the best results in terms of friction reduction obtained by CNTs greases. Steady state tests at a higher average contact pressure of 1.67 GPa proved to have a lower friction coefficient for CNTs grease containing MoS_2; otherwise, CNTs grease without MoS_2 showed an average value of coefficient of friction comparable to commercial calcium and lithium greases, both in a boundary and a mixed regime. Related to the protection against wear, a considerable decrease (−60%) of reference parameter WSD has been measured in the fretting tests with CNTs grease with MoS_2 (NLGI 2) in comparison with the worst conventional grease and −22% in comparison with the best conventional grease. The wear protection increase offered by novel greases was not relevant in the steady state test and further investigations aimed at optimizing such a behavior should be developed. The overall results indicated that the novel carbon nanotube greases show superior tribological properties and will have promising applications in the corresponding industry.

Author Contributions: Conceptualization, A.S. and H.H.; methodology, A.S. and H.H.; software, V.D.; validation, V.D.; investigation, A.S., V.D., H.H. and H.Y; data curation, V.D.; writing—original draft preparation, V.D.; writing—review and editing, A.S., H.H. and H.Y.; supervision, H.H.; funding acquisition, H.H. and A.S. All authors have read and agreed to the published version of the manuscript.

Funding: The authors would like to acknowledge the financial support provided through Army Research Lab (Cooperative agreement W911NF 15-2-0034-S), the South Dakota Advanced Manufacturing Process Technology Transition and Training Center (AMPTECH), University of Salerno "FARB projects".

Conflicts of Interest: The authors declare no conflict of interest.

References

1. Anurag, S.; Prashant, C.; Mamatha, T.G. A review on tribological performance of lubricants with nanoparticles additives. *Mater. Today Proc.* **2020**, *25*, 586–591. [CrossRef]
2. Gulzar, M.; Masjuki, H.H.; Kalam, M.A.; Varman, M.; Zulkifli, N.W.M.; Mufti, R.A.; Zahid, R. Tribological performance of nanoparticles as lubricating oil additives. *J. Nanopart. Res.* **2016**, *18*, 223. [CrossRef]
3. Yadgarov, L.; Petrone, V.; Rosentveig, R.; Feldman, Y.; Tenne, R.; Senatore, A. Tribological studies of rhenium doped fullerene—Like MoS_2 nanoparticles in boundary, mixed and elastohydrodynamic lubrication conditions. *Wear* **2013**, *297*, 1103–1110. [CrossRef]
4. Altavilla, C.; Sarno, M.; Ciambelli, P.; Senatore, A.; Petrone, V. New "chimie douce" approach to the synthesis of hybrid nanosheets of MoS_2 on CNT and their anti-friction and anti-wear properties. *Nanotechnology* **2013**, *24*, 125601. [CrossRef]
5. Wu, H.; Zhao, J.; Cheng, X.; Xia, W.; He, A.; Yun, J.-H.; Huang, S.; Wang, L.; Huang, H.; Jiao, S.; et al. Friction and wear characteristics of TiO_2 nano-additive water-based lubricant on ferritic stainless steel. *Tribol. Int.* **2018**, *117*, 24–38. [CrossRef]
6. Guzman Borda, F.L.; Ribeiro de Oliveira, S.J.; Seabra Monteiro Lazaro, L.M.; Kalab Leiróz, A.J. Experimental investigation of the tribological behavior of lubricants with additive containing copper nanoparticles. *Tribol. Int.* **2018**, *117*, 52–58. [CrossRef]
7. Pape, F.; Pope, G. Investigations on Graphene Platelets as Dry Lubricant and as Grease Additive for Sliding Contacts and Rolling Bearing Application. *Lubricants* **2020**, *8*, 3. [CrossRef]
8. Mello, V.S.; Trajano, M.F.; Guedes, A.E.D.S.; Alves, S.M. Comparison Between the Action of Nano-Oxides and Conventional EP Additives in Boundary Lubrication. *Lubricants* **2020**, *8*, 54. [CrossRef]
9. Xu, J.; Cai, R.; Zhang, Y.; Mu, X. Molybdenum disulfide-based materials with enzyme-like characteristics for biological applications. *Colloids Surf. B Biointerfaces* **2021**, *200*, 111575. [CrossRef]
10. Nabhan, A.; Rashed, A.; Ghazaly, N.M.; Abdo, J.; Haneef, M.D. Tribological Properties of Al_2O_3 Nanoparticles as Lithium Grease Additives. *Lubricants* **2021**, *9*, 9. [CrossRef]

11. Rylski, A.; Siczek, K. The Effect of Addition of Nanoparticles, Especially ZrO$_2$-Based, on Tribological Behavior of Lubricants. *Lubricants* **2020**, *8*, 23. [CrossRef]
12. Morshed, A.; Wu, H.; Jiang, Z. A Comprehensive Review of Water-Based Nanolubricants. *Lubricants* **2021**, *9*, 89. [CrossRef]
13. Coleman, J.N.; Khan, U.; Blau, W.J.; Gun'ko, Y.K. Small but strong: A review of the mechanical properties of carbon nanotube–polymer composites. *Carbon* **2006**, *44*, 1624–1652. [CrossRef]
14. Moghdam, A.D.; Omrani, E.; Menezes, P.L.; Rohatgi, P.K. Mechanical and tribological properties of self-lubricating metal matrix nanocomposites reinforced by carbon nanotubes (CNTs) and graphene—A review. *Compos. Part B* **2015**, *77*, 402–420. [CrossRef]
15. Hong, H.; Thomas, D.; Waynick, A.; Yu, W.; Smith, P.; Roy, W. Carbon nanotube grease with enhanced thermal and electrical conductivities. *J. Nanopart. Res.* **2010**, *12*, 529–535. [CrossRef]
16. Chen, H.; Wei, H.; Chen, M.; Meng, F.; Li, H.; Li, Q. Enhancing the effectiveness of silicone thermal grease by the addition of functionalized carbon nanotubes. *Appl. Surf. Sci.* **2013**, *283*, 525–531. [CrossRef]
17. Yujun, G.; Zhongliang, L.; Guangmeng, Z.; Yanxia, L. Effects of multi-walled carbon nanotubes addition on thermal properties of thermal grease. *Int. J. Heat Mass Transf.* **2014**, *74*, 358–367. [CrossRef]
18. Christensen, G.; Yang, J.; Lou, D.; Hong, G.; Hong, H.; Tolle, C.; Widener, C.; Bailey, C.; Hrabe, R.; Younes, H. Carbon nanotube grease with high electrical conductivity. *Synth. Met.* **2020**, *268*, 116496. [CrossRef]
19. Martin, J.M.; Ohmae, N. Carbon-Based Nanolubricants. In *Nanolubricants*; John Wiley & Sons Ltd.: Chichester, UK, 2008.
20. Curasu, D.L.; Andronescu, C.; Pirva, C.; Ripeanu, R. The efficiency of Co based SWCNT as an AW/EP additive for mineral base oil. *Wear* **2012**, *290–291*, 133–139.
21. Khalil, W.; Mohamed, A.; Bayoumib, M.; Osman, T.A. Tribological properties of dispersed carbon nanotubes in lubricant. *Fuller. Nanotub. Carbon Nanostruct.* **2016**, *24*, 479–485. [CrossRef]
22. Mohamed, A.; Osman, T.A.; Khattab, A.; Zaki, M. Tribological Behaviour of Carbon Nanotubes as Additive on Lithium Grease. *J. Tribol.* **2015**, *137*, 011801. [CrossRef]
23. Kamel, B.M.; Mohamed, A.; Sherbiny, M.E.; Abed, K.A. Tribological behaviour of calcium grease containing carbon nanotubes additives. *Ind. Lubr. Tribol.* **2016**, *68*, 723–728. [CrossRef]
24. Mobasher, A.; Khalil, A.; Khashaba, M.; Osman, T. Effect of MWCNTs/Talc powder nanoparticles on the tribological and thermal conductivity performance of calcium grease. *Ind. Lubr. Tribol.* **2020**, *72*, 9–14. [CrossRef]
25. Mohamed, A.; Ali, S.; Osman, T.A.; Kamel, B.M. Development and manufacturing an automated lubrication machine test for nano grease. *J. Mater. Res. Technol.* **2020**, *9*, 2054–2062. [CrossRef]
26. Mohamed, A.; Tirth, V.; Kamel, B.M. Tribological characterization and rheology of hybrid calcium grease with graphene nanosheets and multi-walled carbon nanotubes as additives. *J. Mater. Res. Technol.* **2020**, *9*, 6178–6185. [CrossRef]
27. Pei, X.; Hu, L.; Liu, W.; Hao, J. Synthesis of water-soluble carbon nanotubes via surface initiated redox polymerization and their tribological properties as water-based lubricant additive. *Eur. Polym. J.* **2008**, *44*, 2458–2464. [CrossRef]
28. Jia, X.; Huang, J.; Li, Y.; Yang, J.; Song, H. Monodisperse Cu nanoparticles@ MoS$_2$ nanosheets as a lubricant additive for improved tribological properties. *Appl. Surf. Sci.* **2019**, *494*, 430–439. [CrossRef]
29. Min, C.; He, Z.; Song, H.; Liu, D.; Jia, W.; Qian, J.; Jin, Y.; Guo, L. Fabrication of novel CeO$_2$/GO/CNTs ternary nanocomposites with enhanced tribological performance. *Appl. Sci.* **2019**, *9*, 170. [CrossRef]
30. Wang, Z.; Ren, R.; Song, H.; Jia, X. Improved tribological properties of the synthesized copper/carbon nanotube nanocomposites for rapeseed oil-based additives. *Appl. Surf. Sci.* **2018**, *428*, 630–639. [CrossRef]
31. Akbarpour, M.R.; Alipour, S.; Safarzadeh, A.; Kim, H.S. Wear and friction behavior of self-lubricating hybrid Cu-(SiC + x CNT) composites. *Compos. B Eng.* **2019**, *158*, 92–101. [CrossRef]
32. Song, W.; Yan, J.; Ji, H. Tribological Study of the SOCNTs@MoS$_2$ Composite as a Lubricant Additive: Synergistic Effect. *Ind. Eng. Chem. Res.* **2018**, *57*, 6878–6887. [CrossRef]
33. ASTM. *Standard D288-61. Standard Definitions of Terms Relating to Petroleum*; ASTM International: West Conshohocken, PA, USA, 1964.
34. Hong, H.; Younes, H.; Christensen, G.; Horton, M.; Qiang, Y. A Rheological Investigation of Carbon Nanotube Grease. *J. Nanosci. Nanotechnol.* **2019**, *19*, 4046–4051. [CrossRef] [PubMed]
35. Hong, H.; Younes, H.; Christensen, G.; Widener, C. Carbon nanotube grease and sustainable manufacturing. *Procedia Manuf.* **2018**, *21*, 623–629. [CrossRef]
36. Liu, H.; Ji, H.; Hong, H.; Younes, H. Tribological properties of carbon nanotube grease. *Ind. Lubr. Tribol.* **2014**, *66*, 579–583. [CrossRef]
37. Younes, H.; Christensen, G.; Groven, L.; Hong, H.; Smith, P. Three dimensional (3D) percolation network structure: Key to form stable carbon nano grease. *J. Appl. Res. Technol.* **2016**, *14*, 375–382. [CrossRef]
38. Christensen, G.; Younes, H.; Hong, G.; Lou, D.; Hong, H.; Widener, C.; Bailey, C.; Hrabe, R. Hydrogen bonding enhanced thermally conductive carbon nano grease. *Synth. Met.* **2020**, *259*, 116213. [CrossRef]
39. Younes, H.; Lou, D.; Hong, H.; Chen, H.; Liu, H.; Qiang, Y. Manufacturable Novel Nanogrease with Superb Physical Properties. *Nanomanuf. Metrol.* **2021**, 1–9. [CrossRef]

Article

Investigation of Tribological Behavior of Lubricating Greases Composed of Different Bio-Based Polymer Thickeners

Seyedmohammad Vafaei [1,*], Dennis Fischer [1], Max Jopen [2], Georg Jacobs [1], Florian König [1] and Ralf Weberskirch [2]

- [1] Institute for Machine Elements and Systems Engineering, RWTH Aachen University, Schinkelstrasse 10, 52062 Aachen, Germany; dennis.fischer@imse.rwth-aachen.de (D.F.); georg.jacobs@imse.rwth-aachen.de (G.J.); florian.koenig@imse.rwth-aachen.de (F.K.)
- [2] Weberskirch Group, Faculty of Chemistry and Chemical Biology, Technical University of Dortmund, Otto-Hahn-Str. 6, 44227 Dortmund, Germany; max.jopen@tu-dortmund.de (M.J.); ralf.weberskirch@tu-dortmund.de (R.W.)
- * Correspondence: seyedmohammad.vafaei@imse.rwth-aachen.de; Tel.: +49-241-809-8260

Abstract: One commonly used lubricant in rolling bearings is grease, which consists of base oil, thickener and small amounts of additives. Commercial greases are mostly produced from petrochemical base oil and thickener. Recently, the development of base oils from renewable resources have been significantly focused on in the lubricant industry. However, to produce an entirely bio-based grease, the thickener must also be produced from renewable materials. Therefore, this work presents the design and evaluation of three different bio-based polymer thickener systems. Tribological tests are performed to characterize lubrication properties of developed bio-based greases. The effect of thickener type on film thickness and friction behavior of the produced bio-based greases is evaluated on a ball-on-disc tribometer. Moreover, the results are compared to a commercial petrochemical grease chosen as benchmark.

Keywords: bio-based grease; grease lubrication; film thickness; friction measurements; polyurea thickener

Citation: Vafaei, S.; Fischer, D.; Jopen, M.; Jacobs, G.; König, F.; Weberskirch, R. Investigation of Tribological Behavior of Lubricating Greases Composed of Different Bio-Based Polymer Thickeners. *Lubricants* **2021**, *9*, 80. https://doi.org/10.3390/lubricants9080080

Received: 21 June 2021
Accepted: 13 August 2021
Published: 17 August 2021

Publisher's Note: MDPI stays neutral with regard to jurisdictional claims in published maps and institutional affiliations.

Copyright: © 2021 by the authors. Licensee MDPI, Basel, Switzerland. This article is an open access article distributed under the terms and conditions of the Creative Commons Attribution (CC BY) license (https://creativecommons.org/licenses/by/4.0/).

1. Introduction

Lubricant industry develops and produces lubricating oils and greases for the lubrication of machine elements such as rolling bearings. Lubricants are essential to reduce friction and thus energy consumption on the one hand, and to prevent premature failure of machine elements due to wear. Lubricants also play an important role in damping the vibrations in bearings and decreasing the dissipated energy by decreasing the friction coefficient in the contact. Most lubricants available on the market today are still produced petrochemically, i.e., on the basis of fossil raw materials. With considering the expected shortage of fossil raw materials and the associated increase in the cost of products based on fossil raw materials, the industry is increasingly focusing on products made from renewable raw materials and the necessary adapted manufacturing processes [1,2]. Furthermore, in Germany, approximately 50% of the mineral oil-based lubricants used are released, each year, to the environment as a result of leakage or accidents, which causes environmental pollution due to chemical materials [3]. The first step toward developing biodegradable lubricants is to design sustainable lubricants, which are made of bio-based materials as an alternative to chemical resources [1].

Bio-based lubricants, such as ester or castor oil, are already available on market, which can be applied as an alternative to mineral oils [4]. In a recent study a bio lubricant is applied in a journal bearing and the friction behavior is compared to a mineral and a synthetic oil. It was found that, the bio lubricant showed lower friction compared to the mineral lubricant, however the synthetic oil showed the lowest friction [5]. In particular, there has been a focus on vegetable oils, such as palm oil and sunflower oil., to find a suitable bio-based lubricant. Vegetable oils in comparison with mineral oils are renewable, easy to obtain, environmentally friendly, economically cheaper and hence, sustainable [6]. There was also a suggestion for application of bio lubricants based on soybean oil improved by epoxy compounds, shown in [7].

Usually, a bearing grease is based on around 70 to 97% of base oil, up to 10% of additives and around 3 to 30 wt.% [8]. According to literature, it is accepted to call a material biogenic or bio-based if it is biodegradable or it is produced partly or completely from renewable raw materials, as defined in [9]. The thickener system can be either based on lithium soaps of fatty acids or petrochemical based oligomers or polymers such as polyureas. Although, bio-based lubricants can be an alternative to commercial petrochemical greases [10], they still need to be developed and a reliable application must be approved. Regarding data from Agency for Renewable Resources (Fachagentur Nachwachsende Rohstoffe e.V.-FNR) the available greases on market are maximum 85% bio-based, since the thickener system is normally not bio-based. To develop a full bio-based grease, base oil and thickener system must be made from renewable materials. Recently, there has been some works on the development of a complete bio-based greases. The results demonstrated in [11] reveal, that a cellulose pulp from the Eucalyptus Globulus plant was obtained as the raw material to produce a thickener system for a bio-based grease. The cellulose pulp was also optimized in another study as a suggestion for bio-based thickener production [12]. Moreover, in [13] it was shown that bio-based greases were developed based on castor oil and high-oleic sunflower oil with cellulose fibers as thickener. In other studies, friction coefficients of developed bio-based greases were comparable to the petrochemical greases [14–16]. Even the wear in bearing with application of bio-based greases were shown to be lower than petrochemical greases in [17]. Although, some bio-based greases have been proposed recently, it becomes more and more difficult to consider economic limits and ecological goals at the concurrently Therefore, new suggestions for bio-based greases are always needed [18]. An alternative for producing bio-based thickeners is to develop polymer-based thickeners, such as polyurea, from sustainable materials, due to their high abilities compared to raw materials. If the thickener system is a lithium soap of conventional high molecular weight fatty acids, it represents a bio-based thickener. The main differences between biogenically thickeners to non-biogenic thickeners is the origin of the raw materials.

Promising bio-based feedstocks for the production of lubricating greases include 1,5-pentamethylene diisocyanate (PDI) [19], 2,5-bisaminomethylenefuran (BAMF), 1,5-pentamethylene diamine (PDA) and 4,4'-diaminodiphenylmethane (MDA). PDA can already be obtained from glucose as reported in [20] and can be reacted with phosgene to give PDI. In this way, PDI with a biocarbon content of up to 70% can be obtained. BAMF can also be synthesized on the basis of glucose [21]. Hydroxymethylfurfural (5-HMF) is first obtained as the base chemical and converted to BAMF in several steps. At the present time, MDA cannot yet be produced bio-based on a large scale. However, MDA is produced on the basis of aniline, the sustainable production of which is already being worked on [22]. Thus, a substitution by bio-MDA is foreseeable in the near future.

Therefore, in this study we present the synthesis of bio-based polyurea thickener systems as well as the production method and chemical characterization of bio-based greases. Moreover, the focus of this study is to elucidate the tribological performance of the produced bio-based greases in comparison to a petrochemical reference grease. To evaluate the tribological performance, the results of film thickness and friction measurements on a ball on disc tribometer are discussed.

2. Lubricants and Methods

In the following the method of development and production of the bio-based greases are explained. Then to characterize the produced bio-based greases, methods of qualifications regarding tribological characteristics of the bio-based greases are explained.

2.1. Bio-Based Grease Development and Production

To produce the polyurea thickener systems, an in-situ polymerization of the monomers in castor oil was carried out. The reaction follows the mechanism of a polyaddition reaction. Three polyureas were prepared based on pentamethylene diisiocyanate (PDI) that was reacted with one of the diamines (a–c) (Scheme 1). As reaction vessel an ULTRA TURRAX® Tube Drive with ST-20 mixing vessel was used. The monomers were heated to melt and mixed in the oil at 100 °C. The PDI was added to the reaction mixture as the last component under stirring. After the addition of all components was completed, stirring was continued for another 2 min. A gel-like structure was typically formed within 10 s. The reaction progress was controlled by FTIR spectroscopy (Bruker Tensor 27, Berlin, Germany) shown in Figure 1. If the conversion was incomplete after two minutes and isocyanate bands were still detectable with the C-N band at 2270 cm^{-1}, more stearyl amine was added to the reaction mixture until the C-N band completely disappeared. All three thickeners have a weight share of 15% of the grease. After cooling, the lubricating greases were homogenized by means of a three-roll mill.

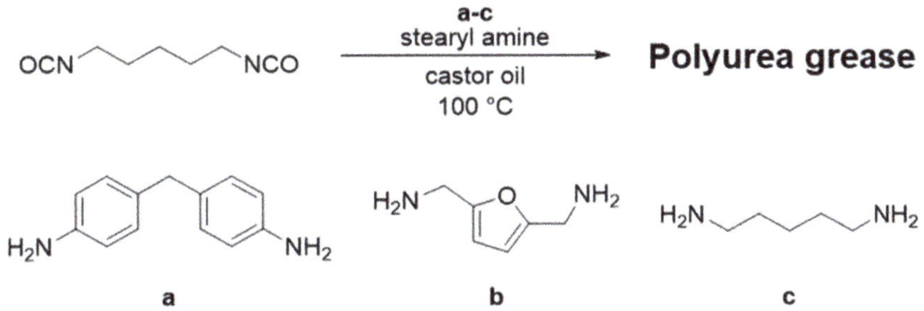

Scheme 1. Reaction scheme of polyurea grease production with all components.

For all three polyurea systems the same monomer ratio was chosen, i.e., diisocyanate:diamine:monoamine was 1:0.75:0.25. This results in a degree of polymerization of 9 for the chain length of the polymer thickeners. The theoretical chain length was calculated using the CAROTHERS equation [23].

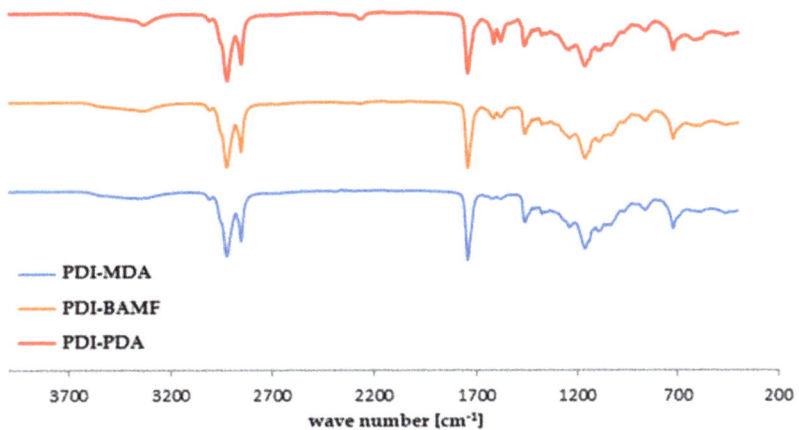

Figure 1. IR-spectroscopy of all three bio-based greases. Bio-based grease 1: PDI-BAMF (orange), bio-based grease 2: PDI-PDA (red) and bio-based grease 3: PDI-MDA (blue).

The successful production of the desired polymeric thickener structures was verified by ^{1}H-NMR spectroscopy (Bruker DPX-400, Berlin, Germany). First, the respective grease was extracted with ethyl acetate in a Soxhlet extractor for three hours. The solid residue was dried under reduced pressure and the thickeners were obtained as yellow powder. Since polyureas are nearly insoluble in organic solvents, the powder was dissolved in concentrated sulfuric acid (98%). To determine the chemical shift δ, a capillary with deuterium oxide phase-separated was used as reference in ^{1}H-NMR spectroscopy.

Using the bio-based grease 1 (PDI-BAMF system) as an example, the analysis of the ^{1}H-NMR spectrum is shown below. Using the ^{1}H-NMR spectrum of the bio-based grease 1 thickener (Figure 2) the successful synthesis of the chemical structure can be confirmed. With ^{1}H-NMR the real number of protons for a structure can be determined, if the amount for one signal is exactly known. As can be seen from Figure 2 with exception of the protons (5) between the urea group and the furan ring, all protons are visible in the spectrum. These protons are overlaid by the signal of the reference D_2O. Since the FTIR spectra (Figure 1) showed that no C-N band is present anymore in the greases, a complete reaction with the termination reagent stearyl amine can be assumed. Thus, there are two alkyl end groups per polymer chain and the protons of the methyl end group (1) can be seen at 0.66 ppm. Since this signal is baseline separated it can be used as an internal reference for the integration of all signals of the spectrum and refers to six protons per polymer chain, due to the fact that each methyl end group has three protons. This makes it possible to determine the exact number of protons for the other signals in the spectrum. For this purpose, the integral of the methyl group protons (1) is set to six. The protons of the CH_2 groups of the alkyl chains (4) in the immediate vicinity of the urea groups also give a baseline separated signal. Therefore, this integral allows an exact determination of the number of repeating units within the polymer chain. For the PDI-BAMF system, four protons must therefore be subtracted from the integral (4), since they belong to the end group. This results in 36 protons. Since there are four protons for (4) per repeat unit, this value is divided by four. This results in an average repeat unit of nine for the polymer chains of the PDI-BAMF thickener. The experimentally determined value of the mean repeating units were in good agreement with the previously calculated using the Carothers equation [23].

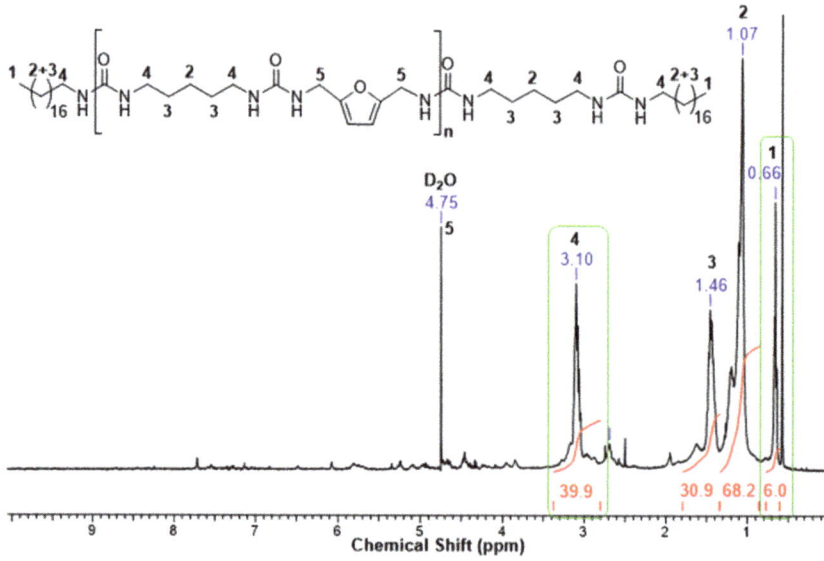

Figure 2. ^{1}H-NMR spectrum of the PDI-BAMF thickener (bio-based grease 1).

Furthermore, the relative flow limit of the three lubricating grease systems was investigated by using an MCR 302 rheometer (Anton Paar, Ostfildern, Germany). Oscillation measurements according to DIN 51810-2 at 25 °C were performed. A plate-plate system with a geometry diameter of 25 mm was used. Forty-one data points were recorded for a shear deformation between 0.01 and 100%. The measurements were performed at a constant angular frequency of 1.59 Hz and a constant gap size of 1.000 mm. The flow limit was determined by the intersection of the logarithmically plotted curves from the storage and loss modulus. It was shown that castor oil is a suitable candidate regarding stability of produced grease and tribological characteristics such as friction coefficient [13,14,24]. Thus, castor oil was chosen as the base oil to produce the bio-based greases.

2.2. Qualification of Bio-Based Greases

The thickener systems and the stable bio-based greases were synthesized at TU Dortmund. The produced bio-based greases and grease properties are listed in Table 1. All bio-based greases had NLGI-class 2 as the result of investigations on penetration depth given in Table A3 in Appendix A. Furthermore, a petrochemical grease is also used for comparison the bio-based greases, which is also listed in Table 1.

Table 1. Summary of the bio-based greases and tested lubricants. PAO = polyalphaolefine.

Grease	Base Oil	Thickener	Base Oil Kinematic Viscosity at 40 °C
Bio-based 1	Castor oil	(a) PDI-BAMF	254 mm^2/s
Bio-based 2	Castor oil	(b) PDI-MDA	254 mm^2/s
Bio-based 3	Castor oil	(c) PDI-PDA	254 mm^2/s
Berutox FH 28 EPK 2	PAO	Polyurea	220 mm^2/s

Tribological tests were performed at the Institute for Machine Elements and Systems Engineering (MSE) of RWTH Aachen University to characterize the greases. For this purpose, the film thickness in the EHD contact was measured on a steel ball-on-glass disc EHD2-Tribometer from PCS Instruments as shown in Figure 3. A polished steel ball with a diameter of 19.05 mm was pushed against a glass disc with 47 N. The resulting Hertzian pressure was aimed to be 700 MPa and the temperature during the measurements was set to 40 °C. The film thickness measurements were conducted based on the interferometry concept, which is described in [25]. This method of measurements was also used in the literature [26]. Film thickness measurements were performed under pure rolling conditions, where the sliding effects were not present in the contact.

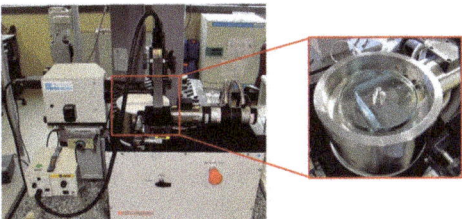

Figure 3. EHD2-ball-disc tribometer of PCS Instruments.

As preparation for the film thickness measurements, the grease was distributed on the disc with a grease film height of approximately 0.1 mm using a grease distributer following the measurement procedure described in previous work by the authors [27]. Before the actual measurement, the grease was uniformly distributed on the rolling track in the contact area by applying a 15 min running-in phase at a constant rolling speed of 30 mm/s. During the film thickness measurement, the rolling speed was continuously increased from 4 mm/s to 1000 mm/s in approximately 45 min. For oils, the relation of film thickness over rolling speeds is known to be linear in double logarithmic diagrams [8].

Therefore, to compare the film thickness of greases over rolling speed to base oil, the film thickness was measured six times at 61 logarithmically distributed rolling speeds under pure rolling condition, i.e., slide-to-roll-ratio (SRR) of 0%.

As the next characterization method, developed bio-based greases were tested by mean of friction losses in the EHD contact. In order to do so, the friction measurements were conducted using the EHD2-ball-on-disc tribometer. In contrast to the film thickness measurements, the steel ball was pressed against a polished steel disc and a torque transducer measured the traction forces acting on the ball. The friction coefficients were calculated by dividing the traction forces by the normal forces given to the ball. The schematic representation of this procedure is shown in Figure 4.

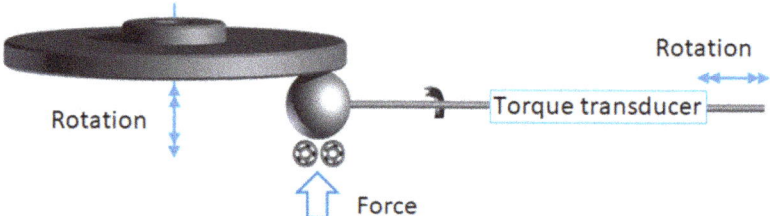

Figure 4. Schematic of the steel ball on steel disc friction measurements on EHD2 tribometer.

The material and roughness properties of the glass disc and steel disc and steel balls for measurements are listed in the Table 2.

Table 2. Table of tribometer disc and ball material properties.

Property	Glass-Disc	Steel-Disc	Steel-Ball
Modulus of Elasticity	75 GPa	207 GPa	207 GPa
Poisson Ratio	0.21	0.29	0.29
Surface Roughness (R_q)	0.8 nm	4.7 nm	6.1 nm [28]

The lubricant conditions in the rolling contacts can be determined based on the definition of lambda ratio as in [29,30]:

$$\lambda = h_0 / \sqrt{\left(R_{q,1}^2 + R_{q,2}^2\right)} \tag{1}$$

where h_0 is the film thickness and R_q is root-mean-square roughness of contact bodies.

According to [29,30], for values of higher than 3 for lambda, fluid regime and separation of both contact bodies is expected. Knowing the roughness of glass disc and steel ball from Table 2, corresponding film thickness value of lambda 3 was calculated equal to 18 nm. Analog to steel all and glass disc, corresponding film thickness of steel ball and steel disc can be calculated as 23 nm. The values show very small film thickness corresponding to lambda = 3. Therefore, it was assumed that no mixed lubrication regime occurs during the film thickness and friction measurements for the whole rolling speed range.

A relative speed between ball and disc was set to induce friction in the contact for friction measurements. Therefore, an SRR of 15% was adjusted. The greases are evenly distributed on the steel disc. Then, a running in phase was set before the actual friction measurement in analogy to the film thickness measurements. The EHD contact pressure for film thickness measurements was also 700 MPa and 40 °C. For each grease, both positive and negative SRR values were measured. This means that both scenarios of faster ball rotation than the disc and vice versa were measured. For each of the positive and negative SRR values, three independent measurements have been performed. In each test the rolling speed was increased from 15 mm/s to 1000 mm/s. The results were averaged to represent the trend of friction coefficient dependent on rolling speed.

3. Results and Discussion

In this section, three subsections are defined to demonstrate the results of characterization of bio-based greases. First, chemical characterization of the greases after production in labor is included. In the second subsection, results of film thickness measurements are explained. The three bio-based greases and the reference grease were compared and evaluated regarding the film formation capability and the occurrence of starvation. In the third subsection, the friction coefficients of the different greases were evaluated dependent on rolling speeds.

3.1. Chemical Characterisation

Through the knowledge of the exact chemical structure of the thickeners, it is possible to determine the proportion of bio-based carbons for all three thickener systems (Table 3). The castor oil used was completely bio-based and accounts for 85 weight percent of the total grease. The remaining 15 weight percent was thickener, as no additives were used. The PDI used has a bio-carbon content of 70%. The BAMF and PDA used were bio-based. Stearyl amine and MDA were used as petrochemical products. For the bio-based grease 1 thickener (PDI-BAMF), this resulted in a value of 64% bio-based carbon. Taking the weight percentages into account, the total value was 95% bio-based carbon. Similarly, for the bio-based grease 3 thickener (PDI-PDA), this resulted in a value of 62 percent bio-based carbon for the thickener and a total value of 94% bio-based carbon by weight for the system. For the bio-based grease 2 thickener (PDI-MDA), only 20% bio-based carbon was obtained for the thickener which results in 88% bio-based carbon by weight. In all three systems, the maximum proportion of bio-based carbon can thus be increased by at least 3%, and in the best case by as much as 10%, compared commercially available systems that are currently used.

Table 3. Overview of all analytical data of the three thickener systems.

Grease	Thickener	Repeating Units	Amount of Bio-Based C [%]	Flow Limit [%]	Dropping Point [°C]
Bio-based 1	(a) PDI-BAMF	9	95	16	226
Bio-based 2	(b) PDI-MDA	9	94	5	193
Bio-based 3	(c) PDI-PDA	9	88	25	252

A flow limit of 16% was determined for the bio-based grease 1. Bio-based grease 3 had a flow limit of 25% shear deformation, which was the highest value of the three systems. Bio-based grease 2 showed the lowest value with a flow limit of only 5% (Figure 5). The flow limit of the bio-based greases can be correlated with the chemical structure of their thickeners. Both bio-based grease 1 and 2 contain aromatic thickener systems. The presence of aromatics in the polymer backbone results in a reduction of the chain mobility and stiffening of the polymer chain that may also facilitate π-π-stacking between different chains [31]. The chain stiffening increases even further for the PDI-MDA system, hence the lowest yield point is expected here. The bio-based grease 3 does not contain aromatics in the polymer backbone of the thickener and thus has the highest chain mobility. This allowed the system to withstand shear stress from the geometry for longer, resulting in a higher flow limit. The correlation of chain mobility as a function of polymer backbone has already been observed in rheology for other applications [32] and is in agreement with the literature.

Additionally, the dropping points of all systems were determined (Table 3). For the bio-based grease 1 a dropping point of 226 °C could be determined. The bio-based grease 3 had a dropping point of 252 °C. For the bio-based grease 2 a dropping point of 193 °C was determined. Class 2 was determined for the NLGI class of all three systems.

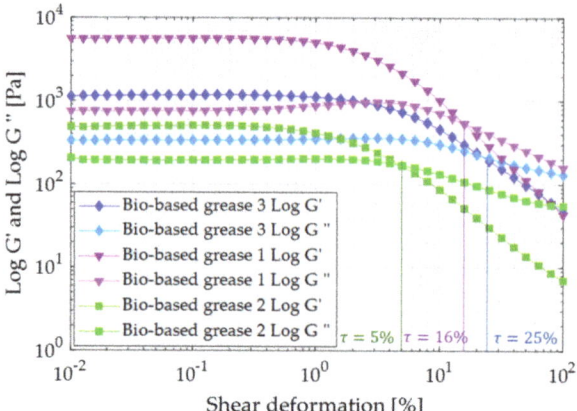

Figure 5. Flow limit determination for bio-based grease 1 (purple), bio-based grease 2 (green) and bio-based grease 3 (blue).

3.2. Film Thickness Measurements

In the first set of tribological characterization we measured the film thickness on ball-on-disc tribometer by varying the rolling speed to identify the onset of starvation for each grease. The results of film thickness measurements are shown in Figure 6.

As a reference, film thickness measurement of castor base oil with a measured kinematic viscosity of 254 mm^2/s at 40 °C is also shown. As stated before, three measurements were averaged to represent the trend of film thickness dependent on rolling speed. The aim of the development of bio-based greases was to identify the bio-based thickener and base oil combinations, which can be used in industry instead of potentially harmful petrochemical based greases and thus reduce harmful environmental influences. Therefore, a petrochemical reference grease was also used to carry out film thickness measurements in comparison to the bio-based greases.

As shown in the upper left diagram in Figure 6, up to the speed of 100 mm/s, film thickness of bio-based grease 1 was higher than base oil. This is caused by thickener effects, as already described in [33–36]. From a rolling speed of 100 mm/s, film thickness increases with further increasing rolling speed, similar to base oil behavior. This effect was the hydrodynamic film formation which mainly depends on the rolling speed and base oil viscosity. In the speed range between 100 mm/s to 1000 mm/s, the base oil dominates the film formation, and thickener effects, assumingly, play a minor role [37]. In slower speeds, up to 100 mm/s, it was shown that bio-based grease 1 had the same film thickness as the reference grease. From 100 mm/s, it was shown that the bio-based grease even higher film thickness than the reference grease, and with increasing the speed the film thickness difference also increases. The hydrodynamic behavior was not captured in reference grease. In upper right diagram in Figure 6, the film thickness of bio-based grease 2 was smaller than base oil curve up to 20 mm/s. The hydrodynamic behavior of bio-based grease 2 ranges from 20 mm/s to 150 mm/s. Above that speed, the film thickness decreases with increasing rolling speed. This may happen due to an insufficient amount of lubricant replenishment in contact [8]. As presented in lower left diagram in Figure 6, hydrodynamic behavior of bio-based grease 3 starts at 20 mm/s and no decrease below the base oil curve can be detected.

In the lower right diagram in Figure 6, three bio-based greases are compared. It can be seen that the bio-based grease 2 shows the smallest film thickness up to 70 mm/s among all bio-based greases. Probably, the effect of thickener in bio-based grease 2 was the least distinctive. Up to 70 mm/s, bio-based grease 1 shows around 100 to 300 nm higher film thickness compared to bio-based grease 3. Thus, bio-based greases 1 and 3 with thickener systems PDI-BAMF and PDI-PDA show a tendency to form higher separating films over the tested range of rolling speeds in comparison to bio-based grease 2 with thickener system PDI-MDA.

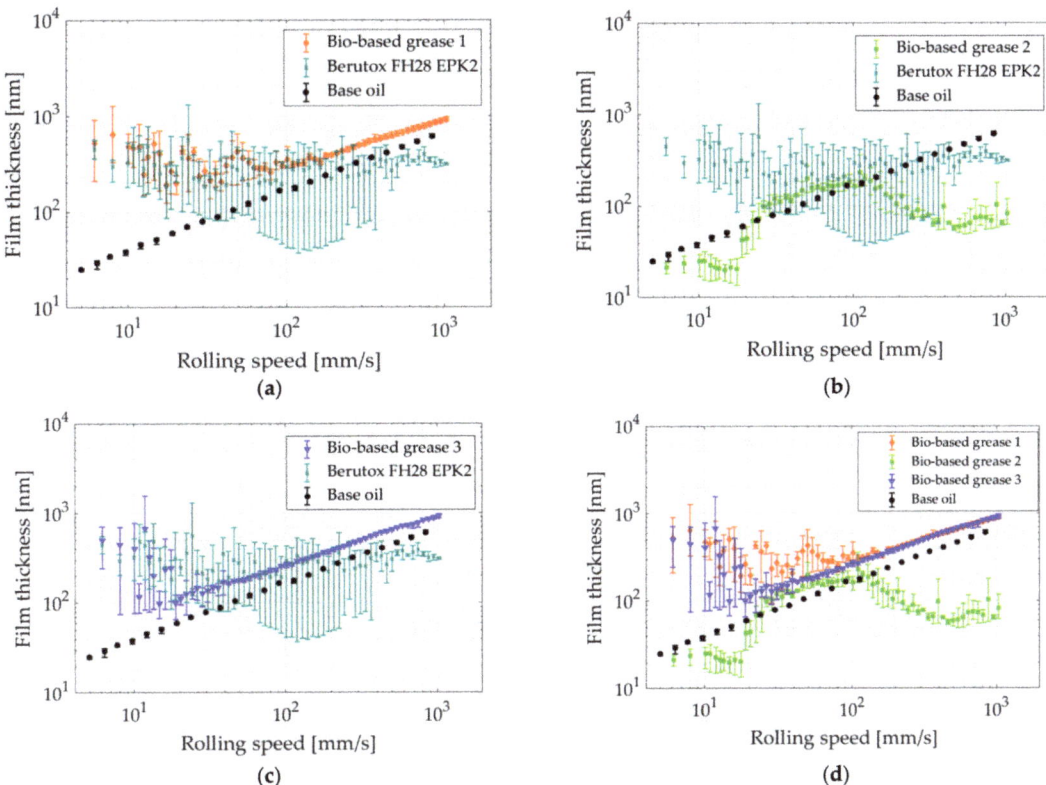

Figure 6. Film thickness of bio-based greases over rolling speed compared to reference grease and base oil at hertzian contact pressure of 700 MPa at 40 °C with SRR = 0 on steel ball on glass disc tribometer: (**a**) film thickness of bio-based grease 1, reference grease and base oil over rolling speed; (**b**) film thickness of bio-based grease 2, reference grease and base oil over rolling speed; (**c**) film thickness of bio-based grease 3, reference grease and base oil over rolling speed; (**d**) film thickness of bio-based grease 1, 2 and 3 and base oil over rolling speed.

The results of the lubricant film thickness measurement also correlate with the chemical structure of the thickeners. As previously shown with the rheology measurements, the chain mobility also has a decisive influence on the formation of the lubricant film. The polymer chains of the PDI-MDA system (bio-based grease 2) have the highest chain stiffness, from which a small contribution of the thickener structure to the formation of the lubricating film can be deduced. The PDI-BAMF system (bio-based grease 1) also contains an aromatic structure, but has a higher chain mobility. The PDI-PDA system (bio-based grease 3) has the highest chain mobility and is, therefore, the most effective in film thickness formation.

3.3. Friction Measurements

The second method of characterization of the bio-based greases in this paper was measuring the produced friction coefficients on EHD contact on steel ball on steel disc measured by EHD2 tribometer by varying the rolling speed. Friction measurements were also conducted on the petrochemical reference grease to have a comparison with the bio-based greases.

The results of friction measurements are shown in Figure 7. In the upper left diagrams of Figure 7, it can be seen that friction coefficient in higher speeds, up to 800 mm/s, was same between bio-based grease 1 and the reference grease. The higher scatter in results was also observed in lower speed due to small grease particles passing through the contact, as was also captured in [38]. As shown in the upper right diagrams of Figure 7, the bio-based grease 2 shows similar friction coefficients to the reference grease up to 30 mm/s. In average speeds, from 30 mm/s to 400 mm/s, friction coefficient of the bio-based grease 1 was lower than the reference grease, and by higher speeds, from 30 mm/s, show both greases again the same friction coefficients. From the lower left diagrams of Figure 7, it can be seen that the friction coefficient of bio-based grease 3 is lower than the reference grease in all speed ranges up to 1000 mm/s. The friction coefficient of bio-based grease 3 decreases even further with increasing rolling speed.

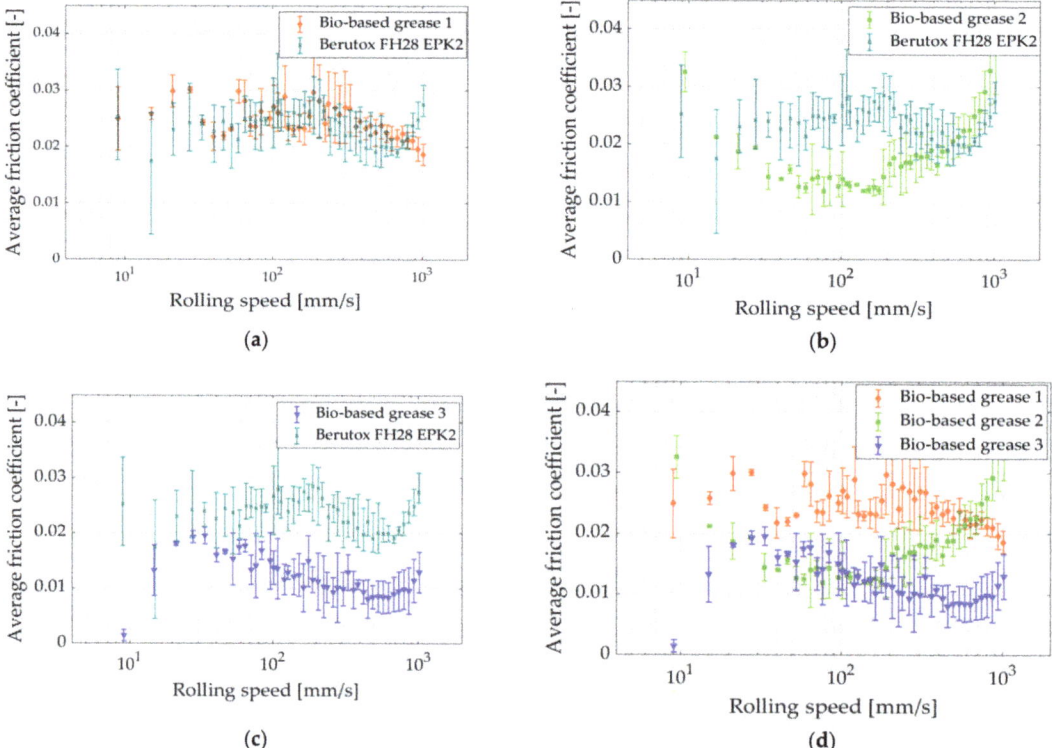

Figure 7. Average friction coefficient of bio-based greases over rolling speed compared to reference grease at hertzian contact pressure of 700 MPa at 40 °C with SRR = 15% on steel ball on steel disc tribometer: (**a**) average friction coefficient of bio-based grease 1 and reference grease over rolling speed; (**b**) average friction coefficient of bio-based grease 2 and reference grease over rolling speed; (**c**) average friction coefficient of bio-based grease 3 and reference grease over rolling speed; (**d**) average friction coefficient of bio-based grease 1, 2 and 3 over rolling speed.

The bio-based greases are compared among each other in lower right diagram in Figure 7. It can be generalized that bio-based grease 3 results in the smallest friction coefficient. Up to 200 mm/s, bio-based grease 2 and 3 lead to similar values. The increase of friction at higher speeds for bio-based grease 3 can be linked directly to the decrease of film thickness and the occurring starvation at those speeds.

4. Conclusions

In this study, bio-based greases based on three bio-based polyurea thickener systems and bio-based castor base oil were developed and the tribological lubrication behavior of the greases was characterized on a ball-on-disc tribometer.

Results of film thickness and friction measurements of the bio-based greases as well as the corresponding pure castor base oil are compared among each other. The main results are summarized as follows:

- The bio-based grease 3 with PDI-PDA and bio-based grease 1 with PDI-BAMF system show comparable chemical and physical properties to known urea thickeners.
- The bio-based grease 2 with PDI-MDA system shows worse chemical and physical properties to known urea thickeners.
- The synthesized thickeners have a higher proportion of bio-based carbon and thus increase the total proportion of bio-based carbon to up to 95% for the whole grease.
- For small velocities up to 100 mm/s rolling speed, bio-based grease 1 shows better film forming behavior due to higher film thickness compared to the other developed greases. In higher speeds show bio-based grease 1 and 3 similar film thickness. The bio-based grease 2 shows smaller film thickness than 1 and 2 in thickener dominant area up to 70 mm/s and in higher speed ranges from 100 mm/s. Bio-based greases 1 and 3 are shown to have comparable to higher film thickness compared to the reference petrochemical grease.
- Friction coefficient of bio-based grease 3 is smaller in comparison to the other investigated greases. Thus, bio-based grease 3 induces less friction in the contact compared to other greases. Friction coefficient of developed bio-based grease 3 is shown to be lower than the comparable petrochemical reference grease.
- Since bio-grease 3 has advantageous film forming and friction behavior, the bio-based thickener type PDI-PDA in combination with castor oil seems to be most promising for full bio-based greases for an application in rolling contacts.

The developed bio-based polyurea greases show promising qualification results compared to reference petrochemical grease regarding film formation and friction coefficients. However, to have a complete bio-based grease, which can be used in industry, further qualification tests such as thermal stability of the greases and anti-wear properties of the greases must be performed. To investigate the anti-wear properties, experiments on bearing test rigs, as shown in [39], must be performed in future works. Furthermore, to develop a fully bio-based grease, production of the bio-based additives also should be focused.

Author Contributions: S.V., D.F. and M.J. wrote the article, designed and performed the experiments and analyzed the data. G.J., F.K. and R.W. supervised the work, discussed the basic design of experiments and provided suggestions for the final discussion. All authors have read and agreed to the published version of the manuscript.

Funding: This research was funded by Agency for Renewable Resources (Fachagentur Nachwachsende Rohstoffe e.V.-FNR) for the project "Entwicklung biobasierter Verdickersysteme zur Herstellung von Schmierfetten".

Institutional Review Board Statement: Not applicable.

Informed Consent Statement: Not applicable.

Data Availability Statement: Supporting data for chemical formulations and characteristics of developed bio-based greases are included in Appendix A.

Acknowledgments: The authors would like to thank kindly all member of the project, Carl Bechem GmbH for supplying the reference grease and Covestro for supplying the PDI.

Conflicts of Interest: The authors declare no conflict of interest.

Appendix A

Castor oil from Carl Roth (Karlsruhe, Germany), 1,5-pentamethylene diisocyanate (PDI) from Covestro (Leverkusen, Germany), 4,4'-diaminodiphenylmethane (a) (MDA), 1,5-pentamethylene diamine (c) (PDA), stearyl amine from TCI (Zwijndrecht, Belgium) and 2,5-Bis(aminomethyl)furan (BAMF) from Carbosynth (Compton, UK) were used to produce the greases (Figures A1 and A2).

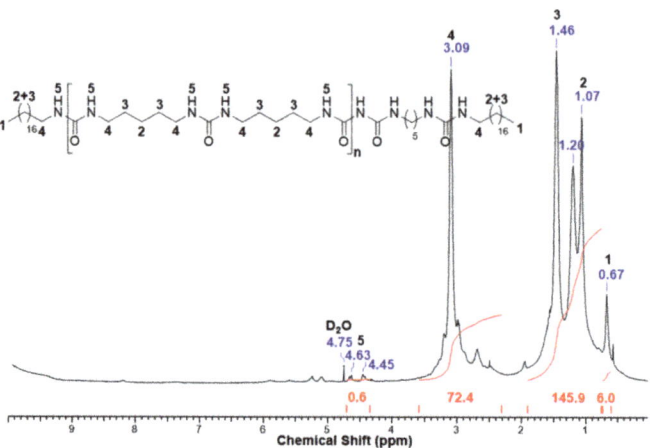

Figure A1. ^{1}H-NMR spectrum of the PDI-PDA thickener.

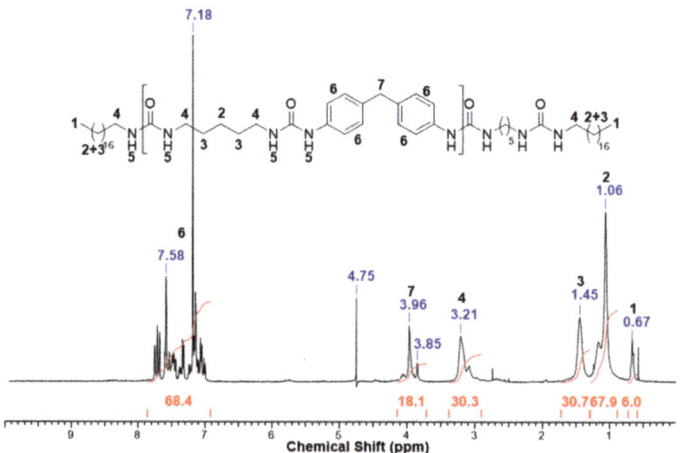

Figure A2. ^{1}H-NMR spectrum of the PDI-MDA thickener.

Table A1. Overview of all analytical data of the three thickener systems.

Thickener	PDI	Diamine	Stearylamine	Castor Oil
(a) PDI-BAMF	483 mg, 1 eq.	296 mg, 0.75 eq.	211 mg, 0.25 eq.	9590 mg
(b) PDI-MDA	398 mg, 1 eq.	384 mg, 0.75 eq.	174 mg, 0.25 eq.	9590 mg
(c) PDI-PDA	520 mg, 1 eq.	258 mg, 0.75 eq.	227 mg, 0.25 eq.	9590 mg

Table A2. NLGI-Class determination.

Bio-Based Grease	Penetration Depth [10^{-1} mm]	NLGI Class
(a) PDI-BAMF	281	2
(b) PDI-MDA	294	2
(c) PDI-PDA	270	2

Table A3. Thermal stability of the bio-based greases.

Bio-Based Grease	Dropping Point [°C]	$T_{degradation\ at\ 10\%\ mass\ lost}$ [°C]
(a) PDI-BAMF	225.8	279
(b) PDI-MDA	192.8	281
(c) PDI-PDA	251.6	275

References

1. Syahir, A.Z.; Zulkifli, N.W.M.; Masjuki, H.H.; Kalam, M.A.; Alabdulkarem, A.; Gulzar, M.; Khuong, L.S.; Harith, M.H. A review on bio-based lubricants and their applications. *J. Clean. Prod.* **2017**, *168*, 997–1016. [CrossRef]
2. Jain, A.; Suhane, A. Capability of Biolubricants as Alternative Lubricant in Industrial and Maintenance Applications. *Int. J. Curr. Eng. Technol.* **2013**, *3*, 179–183.
3. Bundesministerium für Verbraucherschutz, Ernäherung und Landwirtschaft. *Bericht über Biologisch Schnell Abbaubare Schmierstoffe und Hydraulikflüssigkeiten*; Bundesministerium für Verbraucherschutz: Berlin, Germany, 2002.
4. Dwivedi, M.C.; Sapre, S. Total vegetable-oil based greases prepared from castor oil. *J. Synth. Lubr.* **2002**, *19*, 229–241. [CrossRef]
5. Nikolakopoulos, P.; Bompos, D. Experimental Measurements of Journal Bearing Friction Using Mineral, Synthetic, and Bio-Based Lubricants. *Lubricants* **2015**, *3*, 155–163. [CrossRef]
6. Panchal, T.; Chauhan, D.; Thomas, M.; Patel, J. Bio based grease A value added product from renewable resources. *Ind. Crops Prod.* **2015**, *63*, 48–52. [CrossRef]
7. Adhvaryu, A.; Erhan, S.Z. Epoxidized soybean oil as a potential source of high-temperature lubricants. *Ind. Crops Prod.* **2002**, *15*, 247–254. [CrossRef]
8. Lugt, P.M. *Grease Lubrication in Rolling Bearings*; John Wiley & Sons Ltd.: Oxford, UK, 2012; Chapter 9, pp. 213–226.
9. European Commision. Bio-Based Products. Available online: https://ec.europa.eu/growth/sectors/biotechnology/bio-based-products_en (accessed on 9 August 2021).
10. Reeves, C.J.; Siddaiah, A.; Menezes, P.L. A Review on the Science and Technology of Natural and Synthetic Biolubricants. *J. Bio Tribo-Corros.* **2017**, *3*, 1769. [CrossRef]
11. Cortés-Triviño, E.; Valencia, C.; Delgado, M.A.; Franco, J.M. Thermo-rheological and tribological properties of novel bio-lubricating greases thickened with epoxidized lignocellulosic materials. *J. Ind. Eng. Chem.* **2019**, *80*, 626–632. [CrossRef]
12. Martín Alfonso, J.E.; Yañez, R.; Valencia, C.; Franco, J.M.; Díaz, M.J. Optimization of the Methylation Conditions of Kraft Cellulose Pulp for Its Use As a Thickener Agent in Biodegradable Lubricating Greases. *Ind. Eng. Chem. Res.* **2009**, *48*, 6765–6771. [CrossRef]
13. Acar, N.; Kuhn, E.; Franco, J. Tribological and Rheological Characterization of New Completely Biogenic Lubricating Greases: A Comparative Experimental Investigation. *Lubricants* **2018**, *6*, 45. [CrossRef]
14. Sánchez, R.; Franco, J.M.; Kuhn, E.; Fiedler, M. Tribological characterization of green lubricating greases formulated with castor oil and different biogenic thickener agents: A comparative experimental study. *Ind. Lubr. Tribol.* **2011**, 446–452. [CrossRef]
15. Gallego, R.; Arteaga, J.F.; Valencia, C.; Díaz, M.J.; Franco, J.M. Gel-Like Dispersions of HMDI-Cross-Linked Lignocellulosic Materials in Castor Oil: Toward Completely Renewable Lubricating Grease Formulations. *ACS Sustain. Chem. Eng.* **2015**, *3*, 2130–2141. [CrossRef]
16. Nagendramma, P.; Kumar, P. Eco-Friendly Multipurpose Lubricating Greases from Vegetable Residual Oils. *Lubricants* **2015**, *3*, 628–636. [CrossRef]
17. Graça, B.M.; Campos, A.J.V.; Seabra, J.H.O. Taper roller bearings lubricated with bio-greases. *Exp. Mech.* **2009**, *17*, 117–128.
18. Bartz, W.J. Lubricants and the environment. *Tribol. Int.* **1998**, *31*, 35–47. [CrossRef]
19. Morales-Cerrada, R.; Tavernier, R.; Caillol, S. Fully Bio-Based Thermosetting Polyurethanes from Bio-Based Polyols and Iso-cyanates. *Polymers* **2021**, *13*, 1255–1275. [CrossRef]
20. European Commission Directorate-General for Research and Innovation Directorate Bioeconomy. *Bio-Based Products—From Idea to Market—"15 EU Success Stories"*; European Commision: Brussels, Belgium, 2019.
21. Pelckmans, M.; Renders, T.; Van de Vyver, S.; Sels, B.F. Bio-based amines through sustainable heterogeneous catalysis. *Green Chem.* **2017**, *19*, 5303–5331. [CrossRef]
22. Covestro. Bio-Anilin Basis-Chemikalie aus Biomasse. Available online: https://www.covestro.com/de/sustainability/lighthouse-projects/bio-anilin (accessed on 3 May 2021).
23. Carothers, W.H. Polymers and polyfunctionality. *Trans. Faraday Soc.* **1936**, *32*, 39. [CrossRef]

24. El-Adly, R.A.; Bedier, A.H.; Modather, F.H. Jojoba and Castor Oils as Fluid of Biobased Greases: A Comparative Study. *Al-Azhar Bull. Sci.* **2012**, *23*, 29–44.
25. Johnston, G.J.; Wayte, R.; Spikes, H.A. The measurement and study of very thin lubricant films in concentrated contacts. *Tribol. Trans.* **1991**, *34*, 187–194. [CrossRef]
26. Chen, J.; Tanaka, H.; Sugimura, J. Experimental Study of Starvation and Flow Behavior in Grease-Lubricated EHD Contact. *Tribol. Online* **2015**, *10*, 48–55. [CrossRef]
27. Fischer, D.; Jacobs, G.; Stratmann, A.; Burghardt, G. Effect of Base Oil Type in Grease Composition on the Lubricating Film Formation in EHD Contacts. *Lubricants* **2018**, *6*, 32. [CrossRef]
28. Forschungsvereinigung Antriebstechnik e.V. *Dünne Schmierfilme Untersuchung des Schmierfilmaufbaus und der Reibung bei Dünnen Schmierfilmen Mittels Interferometrie und FE8-Wälzlagerversuchen*; IME RWTH Aachen University: Aachen, Germany, 2013.
29. Wittel, H.; Jannasch, D.; Voßiek, J.; Spura, C. *Roloff/Matek Maschinenelemente: Normung, Berechnung, Gestaltung, 23rd revised and expanded ed.*; Springer: Wiesbaden, Germany, 2017.
30. Czichos, H.; Habig, K.-H. *Tribologie-Handbuch, 4th ed*; Springer Fachmedien Wiesbaden: Wiesbaden, Germany, 2015; pp. 127–180.
31. Wheeler, S.E.; Bloom, J.W.G. Toward a More Complete Understanding of Noncovalent Interactions Involving Aromatic Rings. *J. Phys. Chem. A* **2014**, *118*, 6133–6147. [CrossRef]
32. Romo-Uribe, A. On the molecular orientation and viscoelastic behaviour of liquid crystalline polymers: The influence of macromolecular architecture. *Proc. R. Soc. Lond. A* **2001**, *457*, 207–229. [CrossRef]
33. Wilson, A.R. The Relative Thickness of Grease and Oil Films in Rolling Bearings. *Proc. Inst. Mech. Eng.* **1979**, *193*, 185–192. [CrossRef]
34. Cann, P.; Spikes, H. Film Thickness Measurments of Lubrication Greases under Normally Starved Conditions. *NLGI Spokesm.* **1992**, *56*, 21–27.
35. Cann, P.M.E. Thin-film grease lubrication. *Proc. Inst. Mech. Eng. Part J J. Eng. Tribol.* **1999**, *213*, 405–416. [CrossRef]
36. Cann, P.M. Grease lubrication of rolling element bearings—Role of the grease thickener. *Lubr. Sci.* **2007**, *19*, 183–196. [CrossRef]
37. Morales-Espejel, G.E.; Lugt, P.M.; Pasaribu, H.R.; Cen, H. Film thickness in grease lubricated slow rotating rolling bearings. *Tribol. Int.* **2014**, *74*, 7–19. [CrossRef]
38. Cyriac, F.; Lugt, P.M.; Bosman, R.; Padberg, C.J.; Venner, C.H. Effect of Thickener Particle Geometry and Concentration on the Grease EHL Film Thickness at Medium Speeds. *Tribol. Lett.* **2016**, *61*, 470. [CrossRef]
39. Rosenkranz, L.; Richter, S.; Jacobs, G.; Mikitisin, A.; Mayer, J.; Stratmann, A.; König, F. Influence of temperature on wear performance of greases in rolling bearings. *Ind. Lubr. Tribol.* **2021**. [CrossRef]

Article

Effect of Temperature and Surface Roughness on the Tribological Behavior of Electric Motor Greases for Hybrid Bearing Materials

Daniel Sanchez Garrido, Samuel Leventini and Ashlie Martini *

Department of Mechanical Engineering, University of California Merced, 5200 N. Lake Rd., Merced, CA 95343, USA; dsanchezgarrido@ucmerced.edu (D.S.G.); sleventini@ucmerced.edu (S.L.)
* Correspondence: amartini@ucmerced.edu

Abstract: Greased bearings in electric motors (EMs) are subject to a wide range of operational requirements and corresponding micro-environments. Consequently, greases must function effectively in these conditions. Here, the tribological performance of four market-available EM greases was characterized by measuring friction and wear of silicon nitride sliding on hardened 52100 steel. The EM greases evaluated had similar viscosity grades but different combinations of polyurea or lithium thickener with mineral or synthetic base oil. Measurements were performed at a range of temperature and surface roughness conditions to capture behavior in multiple lubrication regimes. Results enabled direct comparison of market-available products across different application-relevant metrics, and the analysis methods developed can be used as a baseline for future studies of EM grease performance.

Keywords: grease; hybrid bearings; electric vehicles; electric motors; friction and wear

Citation: Sanchez Garrido, D.; Leventini, S.; Martini, A. Effect of Temperature and Surface Roughness on the Tribological Behavior of Electric Motor Greases for Hybrid Bearing Materials. *Lubricants* **2021**, *9*, 59. https://doi.org/10.3390/lubricants9060059

Received: 5 April 2021
Accepted: 5 May 2021
Published: 24 May 2021

Publisher's Note: MDPI stays neutral with regard to jurisdictional claims in published maps and institutional affiliations.

Copyright: © 2021 by the authors. Licensee MDPI, Basel, Switzerland. This article is an open access article distributed under the terms and conditions of the Creative Commons Attribution (CC BY) license (https://creativecommons.org/licenses/by/4.0/).

1. Introduction

Electric vehicles (EVs) are emerging as the future of transportation as the market shifts from internal combustion engine vehicles (ICEVs) to electrification [1,2]. Although EVs are more energy efficient than ICEVs, energy losses in electric motors (EMs) are still considerable [3,4]. Mechanical losses in EMs mainly derive from friction in bearings [3,5,6]. Further, about 40–60% of early EM failures are said to be premature bearing faults [4,7], with most failures being due to improper lubrication [7]. Grease is used in 80–90% of rolling bearings [3,8], and, consequently, failed grease lubrication is the predominant cause of EM bearing failure [9–11]. Another source of premature bearing failure that adversely affects EM life is exposure to electrical environments, like those found in EVs [4]. Formulating greases for EV applications is a particular challenge because of key differences between the environment and operating conditions in EVs and those experienced by greases in traditional ICEVs, particularly, speed, temperature, and materials. These conditions are also experienced by greases in industrial EMs.

First, EMs in industrial applications and EVs are operated at high speeds. Grease lubrication is very dependent on speed and can exhibit inverse Stribeck behavior where friction is low at low speeds [12,13]. This deviation from the behavior of lubricating oils is most significant at low λ ratios, i.e., small film thickness to effective surface roughness ratios [12]. In addition, at low speeds or nominal boundary conditions, friction is determined by grease thickener alone and is lower than that predicted for base oil [12,14,15]. Grease film thickness is larger than calculated for a base oil at low speeds, and the same or smaller than calculated at higher speeds [5,16–21]. Therefore, the wide range of speeds expected for EMs introduces additional challenges when selecting or designing greases for EV or industrial applications.

High temperature is another challenge for grease lubrication in EMs. Although some increase in temperature can be beneficial because it helps grease bleed and, thus, resupplies lubricant to the bearing contact track [16,22], high temperatures can also generate harsh operating environments. High rotor speeds generate heat [3], and EMs can reach operating temperatures of 150 °C [23] or even 180 °C [24] for some applications. Therefore, greases used in EM bearings can experience thermal degradation in the form of oxidation and decreased lubricating capabilities as a result of EM operating environments [9–11]. More specifically, grease can suffer thermo-oxidation degradation as a result of high temperature during bearing operation [9], and, at temperatures >120 °C, oxidation ages grease which affects its lubricity and decreases grease life [8,25]. Note that aging will affect the rheological properties of the grease differently depending on its formulation [14]. The result of high temperature and high speed is grease degradation, which affects the chemical composition of the grease, as well as its physical properties that can adversely affect film formation, leading to ineffective lubrication [9].

Another challenge for grease lubrication of EMs is that the materials of bearings may differ from those in traditional motors. Specifically, many EM bearings have ceramic components that act as insulators to mitigate issues related to stray current. Stray current can damage both the bearing and the grease, as well as generate heat, which can cause localized melting of metal surfaces, cause pitting, break particles loose, and embrittle materials [26,27]. In addition, grease that is electrically conductive can amplify these effects and accelerate bearing damage [27]. Current discharge also causes grease degradation by thermal-oxidation and evaporation of the base oil and additives, which then makes grease rigid [26]. The adverse effects of stray current on bearing failure will be more prevalent as EVs become a larger portion of the transportation sector [4]. Therefore, it is necessary to better understand how the electric environment influences EM bearing/grease systems [4] and develop tribological knowledge of non-traditional bearing materials operating in EM environments.

The most common ceramic material used for such applications is silicon nitride. Silicon nitride is suitable for bearings due to its mechanical properties across wide temperature ranges, electrical insulation, thermal shock resistance, excellent fracture toughness, wear resistance, long life, and reliable low maintenance operation [28,29]. Silicon nitride serves as a bearing insulator [26] that disrupts stray current in EMs and, thus, minimizes grease thermal degradation and melting of material that leads to wear. Often, silicon nitride is used in hybrid bearings that consist of ceramic rolling elements and traditional steel raceways [26]. Hybrid bearings have been found to last longer than predicted based on the Lundberg-Palmgren theory [30], and grease life with hybrid bearings was found to be up to four times longer than with traditional all steel bearings [26]. However, there are issues associated with the use of ceramic bearing elements, particularly related to lubricant additives. For example, phosphorus-based additives were found to not react with silicon nitride as they would with steel so the tribofilms formed were not effective in improving silicon nitride bearing life [31–33]. Another potential issue is that hybrid bearings experience higher contact stress than all steel bearings under the same applied load [30], due to the higher hardness of ceramics, which corresponds to smaller contact areas.

The aforementioned challenges with grease lubrication in EMs can be partially addressed through design or selection of greases specifically for EM environments. Thickener type, base oil type, and viscosity all have a significant impact on film thickness and friction for grease lubricated rolling/sliding contacts [12–14,34]. Greases are continually changing, as formulators and designers seek to optimize lubrication in different operating environments, while remaining compatible with component materials. For example, recent studies have used nanotechnology to create novel additives for grease formulations that improve lubricity and grease life [3,35–37]. Another study focused on extending EM bearing life by reducing grease degradation and found this can be achieved with the use of an antioxidant and high-temperature composite grease formulation [9]. Continued research in the area of grease formulation is crucial because grease behavior is extremely application dependent [22].

Grease optimization often focuses on identifying the best combination of base oil and thickener for EM applications. Studies have evaluated the lubrication mechanisms associated with synthetic or mineral base oil with urea or lithium thickener. For instance, an ester-polyurea grease used for EMs with silicon nitride ceramic rolling elements was found to have excellent life, resist high operating temperatures, and withstand high speeds experienced in the motor [26]. Synthetic base oils resist higher temperatures, while generating low friction and improving service life [3,38]. Although lithium thickened greases are currently the most widely used [8], some studies suggest urea thickened grease may generate lower friction and thicker films, as well as have a closer correlation to typical Stribeck behavior than lithium thickened greases [12]. A study comparing custom polyurea and lithium thickened greases on bearing steel was performed to characterize performance at 25, 70, and 120 °C and average surface roughness of 10, 100, and 200 nm [12]. Polyurea greases were shown to have the lowest friction at low speed, average surface roughness of 100 nm Ra, and temperatures of 70 and 120 °C. Further, it was reported that polyurea had thicker low-speed films than lithium greases [12].

Based on the current findings, it is evident that both base oil and thickener affect grease performance and that optimizing this performance for EM applications requires characterization at the conditions in which the motor will operate. Specifically, EM greases are subject to higher temperatures and may be required to function with different bearing materials than traditional applications. Here, we tested the tribological performance of four commercially available greases with formulations/additives designed for EM applications, with different combinations of mineral or synthetic base oil with lithium (complex) or urea thickeners. The study focused on lubrication of silicon nitride sliding on steel across a range of temperature and surface roughness conditions. The tribological performance of these greases and bearing materials was quantified in terms of friction and wear. Characterization included both ball-on-disk and 4-ball tests, as well as an analysis of the results in terms of lubrication regimes. Finally, the four greases were evaluated based on a ranking system that emphasized priorities for EM applications.

2. Materials and Methods

Four commercially available greases were studied, all of which were designed for EM applications, per manufacture specifications. All EM greases had an International Standard Organization Viscosity Grade (ISO VG) of 100 and a National Lubricating Grease Institute (NLGI) grade of 2, but with different combinations of thickener and base oil types. The specific greases studied were: synthetic-polyurea (SP), mineral-polyurea (MP), mineral-lithium (ML), and synthetic-lithium complex (SL). Note that, since the tested greases were commercially available, formulation details, such as additive composition and concentration, were not known. Table 1 provides the reported information for each grease.

Table 1. EM grease specifications.

EM Grease	Acronym	Base Oil Viscosity at 40 °C (cSt)	Base Oil Viscosity at 100 °C (cSt)	Base Oil Density at 15 °C (g/cm^3)	Dropping Point (ASTM D2265 °C)
Synthetic-polyurea	SP	100	14	0.85	250
Mineral-polyurea	MP	100	12	0.88	260
Mineral-lithium	ML	100	11	0.93	180
Synthetic-lithium complex	SL	100	14	0.85	260

Two types of experiments were conducted to characterize the tribological performance of the EM greases. First, a Rtec Instruments Multi-Function Tribometer equipped with a temperature chamber was used to perform unidirectional sliding ball-on-disk tests, illustrated in Figure 1a. In those tests, a silicon nitride ceramic bearing ball with a 9.525 mm diameter and an average surface roughness (Ra) of 20 nm were used. The ceramic balls met grade 5 quality specifications. The flat disk had a 50.8 mm diameter and was made of hardened 52100 steel. An Allied High Tech Metprep 3 polisher was used to polish the disks with the use of a silicon carbide abrasive pad in water suspension for non-directional surface finish. The disks were polished to achieve a final average surface roughness of 10, 35, 60, 120, or 200 nm. Based on the ball and disk roughness, average composite roughness cases evaluated were 22, 40, 63, 122, and 201 nm. The surface roughness of the disks was measured using a Bruker DektakXT contact mode profilometer. Prior to testing, all testing surfaces were ultrasonically cleaned in heptane.

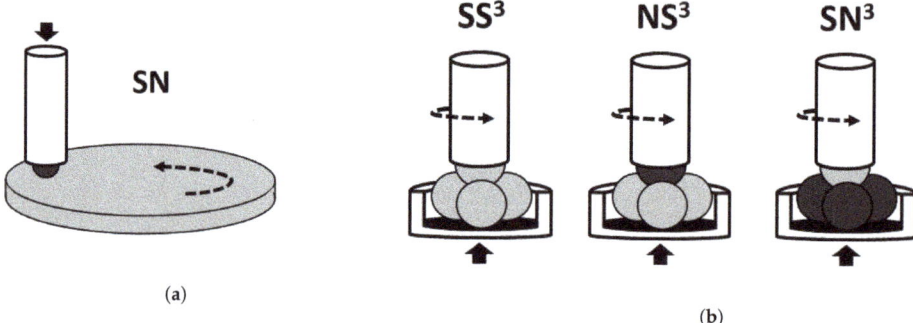

Figure 1. Greases were characterized using two test configurations: (**a**) Ball-on-disk and (**b**) 4-ball. The tests were performed with various combinations of steel (S) and silicon nitride ceramic (N) samples.

Ball-on-disk test parameters closely adhered to ASTM D5707-16 with some modifications to capture key conditions expected in EM environments, specifically, bearing material, temperature and surface roughness. The load was 10 N, corresponding to a maximum Hertz contact pressure of 1.2 GPa, and the sliding speed was 250 mm/s. Temperatures tested were 40, 100, and 150 °C. For each test, about 500 mm^3 (pea size amount) of grease was applied to lubricate the samples. A grease scoop was employed for tests at 40 °C to avoid starvation by continuously pushing the grease back onto the track [12,13,17,18]. All tests were run to 400 m total sliding distance, and each test condition was repeated three times.

Second, a Falex Multi-Specimen Test Machine was used to perform 4-ball testing, illustrated in Figure 1b. The 4-ball tests enabled different combinations of silicon nitride and hardened 52100 steel to be evaluated. Three different material configurations were tested:

- one steel rotating element on three steel stationary elements (SS^3),
- one silicon nitride ceramic rotating element on three steel stationary elements (NS^3), and
- one steel rotating element on three silicon nitride ceramic stationary elements (SN^3).

The SS^3 case resembled a traditional all steel bearing assembly, and the NS^3 case resembled a hybrid bearing assembly. The SN^3 resembled an inverted hybrid bearing assembly, meaning that the material typically used for the races was used as the rolling element and vice versa. All-ceramic bearings are infrequently used for standard applications, so the NN^3 configuration was not tested here. Test parameters followed the ASTM-D2266 standard. The load was 392 ± 2 N, which corresponds to a maximum Hertz contact pressure of 4.6 GPa for the SS^3 configuration and 5.2 GPa for the NS^3/SN^3 configurations. The silicon nitride ceramic test balls met grade 5 quality specifications and had a surface roughness of 20 nm Ra. The steel balls met grade 10 specifications and had 25 nm Ra. The speed was 1200 ± 60 revolutions per minute for 60 minutes, and the temperature was held at 75 °C ± 2 °C. All four greases were tested using each material configuration; the SS^3 and NS^3 results are averages of three tests, and the SN^3 results are averages of two tests.

Error was calculated as the difference between the maximum and minimum wear for each configuration and grease.

Images captured of the worn ball surfaces were obtained using a Leica Optical Microscope (Model DM 2500M) for both test methods. For the ball-on-disk tests, wear volume was calculated per ASTM G-133-05. Specific wear rate was then calculated by dividing the volume by the load and total sliding distance. For the 4-ball test, wear scar diameters were measured per ASTM D2266-01. Those measurements were then used to calculate the wear area of the scar.

3. Results
3.1. Ball-on-Disk Wear

Wear rate as a function of surface roughness for all EM greases is shown in Figure 2a. For smooth surfaces, the wear rates of the four greases are similar, although the lowest wear is observed for the MP. In contrast, there is more differentiation between the greases on rougher surfaces, where the ML consistently exhibits the lowest wear rate. In addition, for these testing parameters, greases with mineral base oil have lower wear rate than the synthetic base greases.

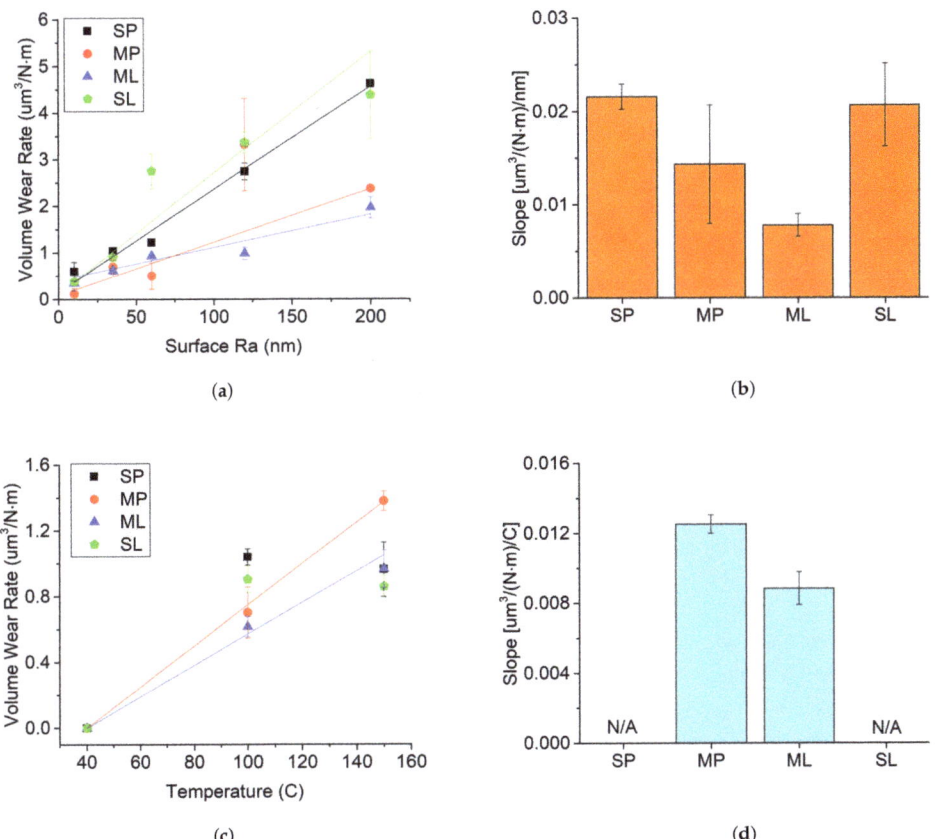

Figure 2. Wear results from EM grease ball-on-disk tests. (**a**) Wear rate as a function of roughness and (**b**) change in wear rate with roughness at 100 °C. (**c**) Wear rate as a function of temperature and (**d**) the change in wear rate with temperature at 35 nm Ra (composite Ra of 40 nm) for MP and ML.

The sensitivity of wear rate to changes in roughness was quantified as the slope of a linear fit to the data. Although this analysis is based on an assumption that wear rate increases linearly with roughness, the approach enables direct comparison of the greases. The slope calculated from a linear fit to the wear rate versus roughness data is shown in Figure 2b. This analysis indicates that wear rate with the ML grease is the least dependent on surface roughness. In addition, of the tested greases, greases with mineral base oil have less wear-roughness dependence than the synthetic base greases.

Wear rates at different temperatures are shown in Figure 2c. At 40 °C, there is no observable wear for any of the greases. The lowest wear rate at 100 °C is observed for the ML grease and, at 150 °C, is found for the SL grease. Additionally, at 100 and 150 °C, for the greases tested here, lithium thickened greases have a lower wear rate than their polyurea counterparts.

The temperature dependence of the wear rate is very different for synthetic versus mineral based greases. Specifically, the wear rate increases approximately linearly between 100 and 150 °C for the mineral greases, but is nearly constant for the synthetics. Due to this behavior, the linear approximation cannot be used to quantify the change of wear rate with temperature for the synthetic greases. However, the linear fit was performed for the mineral greases as shown in Figure 2d. The wear rate is less dependent on temperature for the ML grease than the MP grease.

3.2. Ball-on-Disk Friction

Friction results for each grease are shown in Figure 3. On average, friction increased with surface roughness for all greases (see Figure 3a). In addition, on most surfaces, friction was lowest for the SL grease. On the rougher surfaces, the ML also exhibited low friction behavior. For these tests, the lithium based greases had lower friction than the polyurea greases, except on the smoothest surfaces where the friction coefficient was below 0.08 for all greases.

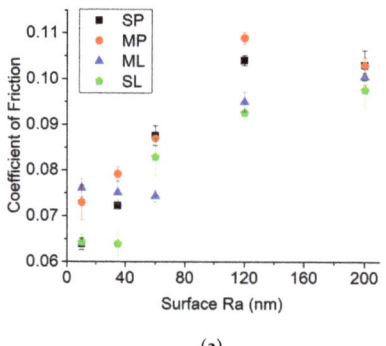
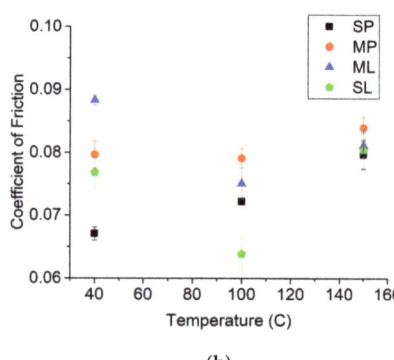

(a) (b)

Figure 3. Friction results from EM grease ball-on-disk tests. Friction coefficient (**a**) as a function of surface roughness at 100 °C and (**b**) as a function of temperature at 35 nm Ra (composite Ra of 40 nm).

Friction at three different temperatures is shown in Figure 3b. At 40 °C, the lowest friction was exhibited by the SP grease whereas, at 100 °C, the SL grease had the lowest friction. At both 40 and 100 °C, the friction was lower for synthetic greases than their mineral counterparts. At 150 °C, the friction coefficient was comparable for all four greases.

The friction trends with respect to roughness and temperature are not linear. This is primarily because both roughness and temperature affect the lubrication regime. Further, increasing temperature can promote grease bleed such that the degree of starvation decreases with increasing temperature [16]. So, the effect of these parameters on friction cannot be quantified using a simple linear fit. Instead, these trends will be analyzed in the context of the Stribeck curve, as discussed later.

3.3. 4-Ball Test

Results from the 4-ball tests are shown in Figure 4. ML had the lowest wear across all three bearing configurations. The performance of ML might be attributable to thicker lubricating films that provide more separation between interacting surfaces or better anti-wear film formation. For the SS^3 and NS^3 configurations, average wear increased as ML < MP < SP < SL. However, this trend cannot be directly explained since the 4-ball test is primarily measuring anti-wear behavior, and the additive composition of these commercial greases is unknown. For the SN^3 configuration, wear was high for all four greases, and large error bars precluded direct comparison between the greases.

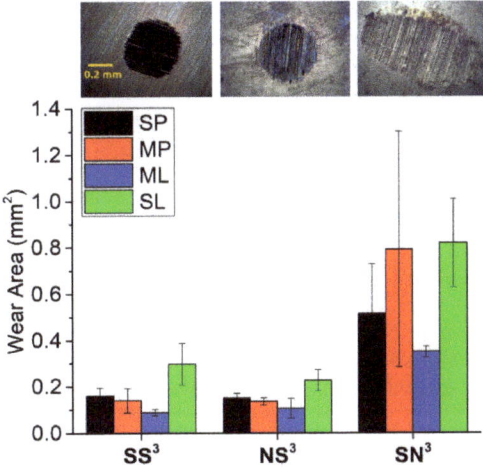

Figure 4. Wear area for four greases and three bearing configurations measured using the 4-ball test. Representative wear patterns (from left to right): SS^3 circular wear scar on steel ball, NS^3 circular wear scar on steel ball, and SN^3 elliptical wear scar on ceramic ball.

Comparing the different material combinations, for all greases, the lowest average wear was observed for NS^3, followed by SS^3 and then SN^3. The observation that wear for NS^3 was lower than that for SS^3 is consistent with previous reports that grease life with hybrid bearings is longer than with standard bearings [26]. Lower wear for NS^3 is also consistent with experimental and anecdotal observations that suggest longer lives for hybrid bearings than estimated by the Lundberg-Palmgren equations [30].

In contrast, the SN^3 configuration consistently had very high wear. This configuration also exhibited qualitatively very different behavior than the other two material pairs. As shown in the insets to Figure 4, the wear scars for the SS^3 and NS^3 configurations are circular, while those for the SN^3 are elliptical. The wear mechanism of the rotating elements determine and may cause non-circular wear scars of the stationary balls [39]. Therefore, the difference may be attributable to the hardness of the rotating element. For SN^3, the steel ball is the rotating element attached to the spindle (upper ball), while the three lower balls are silicon nitride. Material hardness affects material wear; a softer steel ball rotating on a harder ceramic ball causes the wear scar to elongate with increasing material deformation, thus causing relative displacement between the upper and lower balls.

4. Analysis & Discussion

4.1. Lubrication Regime Analysis

The friction results shown in Figure 3 suggests that changing either roughness or temperature caused a transition between lubrication regimes. The lubrication regime can be determined by the lambda ratio:

$$\lambda = \frac{h}{\left(Ra_{ball}^2 + Ra_{disk}^2\right)^{1/2}}, \quad (1)$$

where h is the film thickness, Ra_{ball} is the average roughness of the ball, and Ra_{disk} is the average roughness of the disk. Although the exact values of λ corresponding to transitions between lubrication regimes vary in the literature, they are often defined by $\lambda \gtrsim 3$ for full film lubrication, $1 \lesssim \lambda \lesssim 3$ for mixed lubrication, and $\lambda \lesssim 1$ for boundary lubrication [34,40,41]. However, these transition values are not absolute, and studies have shown that full film or mixed lubrication is possible even in cases where λ would typically suggest boundary lubrication (e.g., $\lambda \lesssim 1$) [41].

For bearings, the λ ratio also affects contact fatigue life. Low λ ratios are associated with surface deformation and distress [2]. In the context of the conditions studied here, small surface roughness and low temperature conditions that correspond to higher λ ratios will have lower contact fatigue and longer life.

To calculate λ, we first had to determine film thickness. It is known that the film thickness of a grease may be larger or smaller than the film thickness for its base oil, depending on the operating conditions [12,34]. However, there is no standard equation or method of calculating grease film thickness that is applicable for all conditions. Therefore, as a first order approximation, we calculated film thickness using the Hamrock and Dowson equation [2] for central film thickness with parameters for the base oil:

$$h \approx h_c = 2.69 R \left(\frac{U\eta}{ER}\right)^{0.67} (\alpha E)^{0.53} \left(\frac{W}{ER^2}\right)^{-0.067} (1 - 0.61 e^{-0.73k}), \quad (2)$$

where U is the speed, R is effective radius, E is effective elastic modulus, α is the pressure-viscosity coefficient, η is the ambient viscosity, W is the load, and $k = 1$ for a spherical geometry. Most of these parameters are constant for the ball-on-disk tests. However, the ambient viscosity and pressure-viscosity coefficient were calculated for each test based on the rheological properties of the base oil and the temperature. Table 2 summarizes the film thickness and λ ratio for each EM grease, temperature, and roughness case considered in this study.

The friction measured from the ball-on-disk tests is plotted as a function of the calculated λ ratio to create a Stribeck curve in Figure 5. The large λ cases correspond to tests run on smooth surfaces and at lower temperatures. Conversely, rough surfaces and high temperatures lead to small λ ratios. The general shape of the Stribeck curve in Figure 5 indicates that our tests included the mixed regime, where friction decreases with λ, and the full film regime, where friction increases with λ.

Table 2. EM grease calculated film thickness (h_c in nm) and lambda (λ) ratio at all tested composite roughness and temperature combinations.

Temperature	°C	40	100	100	100	100	100	150
Composite Roughness	nm	40.3	22.4	40.3	63.3	121.7	201	40.3
SP	h_c	117	28.3	28.3	28.3	28.3	28.3	13.7
	λ	2.89	1.27	0.70	0.45	0.23	0.14	0.34
MP	h_c	151	27.8	27.8	27.8	27.8	27.8	11.3
	λ	3.74	1.25	0.69	0.44	0.23	0.14	0.28
ML	h_c	142	26.9	26.9	26.9	26.9	26.9	11.0
	λ	3.52	1.20	0.67	0.42	0.22	0.13	0.27
SL	h_c	117	28.3	28.3	28.3	28.3	28.3	13.7
	λ	2.89	1.27	0.70	0.45	0.23	0.14	0.34

The greases clearly exhibit full film at larger lambda ratios. In this regime, the lowest friction was exhibited by the SP and SL (synthetic greases). The mixed regime is clearly observed at small λ ratios. Here, as composite roughness increases, λ values decrease, and friction tends to increase. In mixed lubrication, the lowest friction was observed for the ML and SL (greases with lithium thickener). Across most of the lubrication regimes measured, SL had the best friction performance.

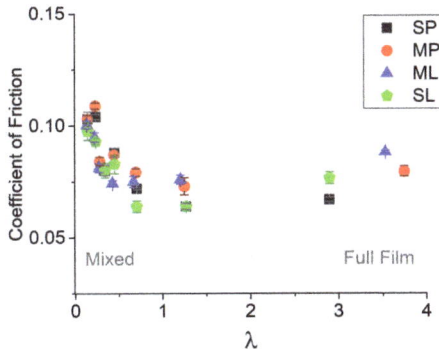

Figure 5. Stribeck curve based on measured friction and calculated λ ratios for four greases tested across all roughness and temperature conditions.

The transition between the full film and mixed lubrication regimes is important because both friction and wear are higher in the mixed regime due to asperity contacts in the interface. Therefore, it is desirable to remain in the full film regime as long as possible. To identify the λ ratio at which the full film-mixed transition occurs for each grease, we found the intersection of a linear fit to the data in the mixed regime and a linear fit to the data in the full film regime. The two largest λ ratios for each grease were fit for full film, and the three smallest λ ratios were fit for the mixed regime. Figure 6 shows the linear fits and their intersection which was identified as the transition lambda (λ_t). The λ_t values for each grease were found to be: SP at $\lambda_t = 0.48$, MP at $\lambda_t = 0.47$, ML at $\lambda_t = 0.37$, and SL at $\lambda_t = 0.58$.

The ML grease had the lowest λ_t, indicating that the interface would remain in the full film regime the longest with increasing temperature or roughness. However, it is important to note that ML also has higher friction in this transition region. So, ML's lower λ_t suggests the lubricant is able to maintain a thicker lubrication film than the other greases, but this comes at a cost of higher viscous friction. On the other hand, SL has a larger λ_t value but considerably lower friction than the rest of the tested greases in this transition range. In fact, despite having a larger λ_t value, SL maintained lower friction at most test conditions. This analysis shows there is a compromise between low friction in full film lubrication and how long the interface will remain in that regime before the onset of mixed lubrication.

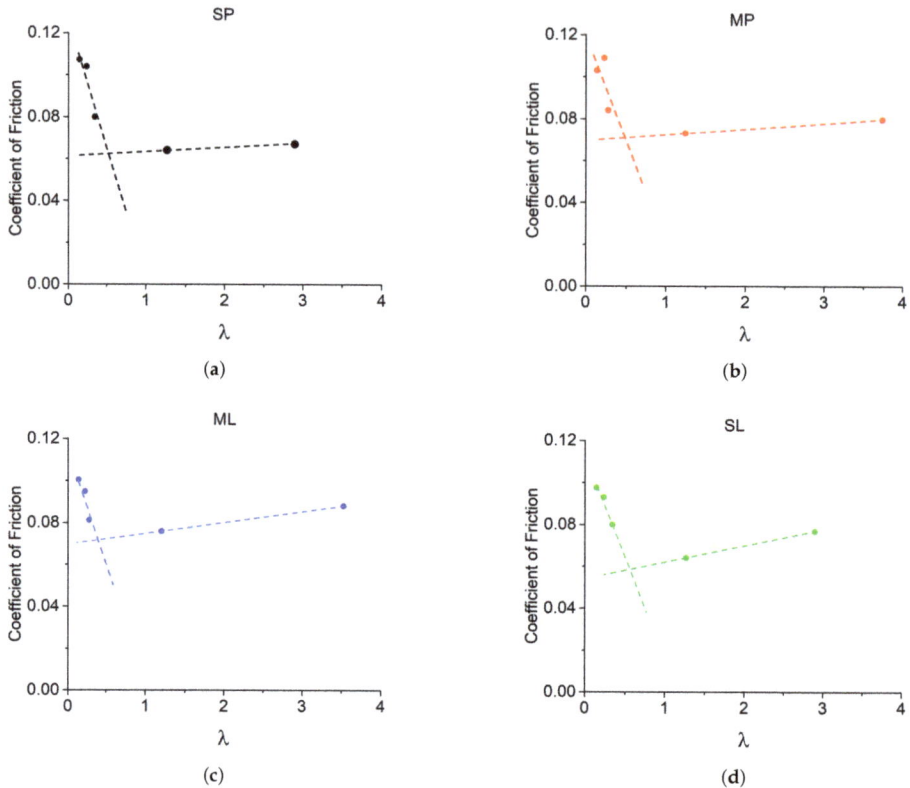

Figure 6. Independent linear fits performed for the mixed and full film regimes for (**a**) SP, (**b**) MP, (**c**) ML, (**d**) SL. The intersection of the two lines corresponds to the transition lambda λ_t.

4.2. Predicted Lubrication Regime Transitions

The λ ratio determines lubrication regime, as well as contact fatigue. In our study, this critical ratio is determined by surface roughness, grease properties and temperature. So, for a given grease, roughness, and temperature, the λ value can be calculated, and the conditions at which the lubrication regime transitions to mixed can be predicted.

Surface roughness affects this calculation directly, as it appears in the denominator of Equation (1). Temperature indirectly affects the film thickness as calculated using Equation (2) through its effect on η and α. The grease itself determines the values of η and α and their temperature dependence. To predict λ for any temperature T, we used a linear equation for $\alpha(T)$ and the Vogel equation [2] for $\eta(T)$, both fit to available grease data. Then, λ was calculated directly using the equations for $\alpha(T)$ and $\eta(T)$, combined with Equations (1) and (2).

This analysis was performed for each grease at temperatures ranging from 30 to 200 °C and composite roughness values from 20 to 200 nm Ra. The predicted λ values are shown as color contour plots in Figure 7. In addition, shown are horizontal planes corresponding to λ_t, the ratio at the transition between mixed and full film lubrication, identified from the friction data for each grease in Figure 6. The intersection between this plane and the surface predicted as described above indicates the temperature and surface roughness at which the interface will transition from full film to mixed lubrication. Such an approach can be used as part of the design process to guide selection of a grease, surface roughness specifications, or prescribed limits on operating conditions.

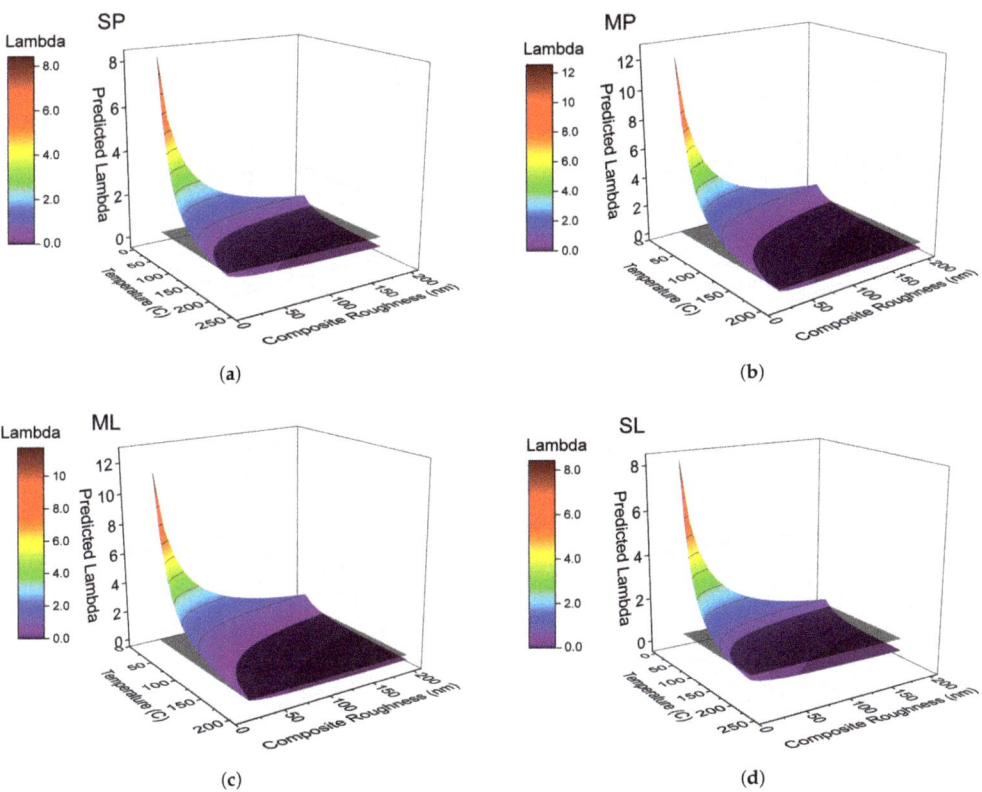

Figure 7. Contour plots with predicted λ ratios for each of the four commercially available EM greases: (**a**) SP, (**b**) MP, (**c**) ML, and (**d**) SL. The transition between full film and mixed lubrication (λ_t) is shown as a horizontal plane.

4.3. Grease Evaluation

The four greases evaluated in this study exhibited varying levels of performance at different surface roughness and temperature conditions. These observations are summarized briefly here.

As observed in Figure 2a, MP had the lowest wear rate on smooth surfaces, while ML had the lowest wear on rough surfaces. ML was also found to exhibit the least dependence of wear rate on surface roughness. On average, greases with mineral base oil had lower wear rate and roughness dependence than the synthetic base greases. In high temperature tests (150 °C), the lowest wear rate was found for the SL grease. In addition, the wear rate of the synthetic greases did not change with temperature at high temperatures, while an increase in wear rate with temperature was observed for the mineral greases. In the 4-ball tests, ML had the lowest wear across all the various bearing configurations. For the SS^3 and NS^3 configurations, average wear increased as ML < MP < SP < SL. Larger wear rates increase material debris, which can have implications, such as artificial surface roughness, reduced lubricating capabilities, and abrasion/erosion, so low wear is extremely important.

In terms of friction, on most surfaces, friction was lowest for the SL grease. For rougher surfaces, the lithium greases generally had lower friction than the polyurea greases. On very smooth surfaces, the synthetics had lower friction than the mineral based greases. In terms of temperature, at 40 °C, the lowest friction was exhibited by the SP grease whereas, at 100 °C, the SL grease had the lowest friction. At both 40 and 100 °C, the friction was lower for synthetic greases than their mineral counterparts. The friction data was also used to determine transitions between the full film and mixed lubrication regime. This analysis showed that under these testing parameters, ML had the lowest λ_t, indicating that the interface would remain in the full film regime the longest with increasing temperature or roughness. However, ML also had higher friction in this transition region indicating that the lubricant maintained a thicker lubrication film at a cost of higher viscous friction. In contrast, SL had a larger λ_t ratio but considerably lower friction than the rest of the tested greases in this range.

While the comparisons between greases in terms of individual performance metrics are valuable, they need to be combined to determine which grease is best for a given application. Therefore, it is necessary to develop a grease evaluation and comparison method to assess these commercially available greases. Performance metrics included are low temperature (40 °C) friction, low surface roughness (10–60 nm Ra) friction and wear, high surface roughness (120–200 nm Ra) friction and wear, high temperature (100–150 °C) friction and wear, wear dependence on surface roughness, and NS^3 (best represents EM hybrid bearings) wear from the 4-ball tests. The ranking system was developed with EM bearing applications in mind, so high temperature friction and wear were given twice the weight of the other metrics. The greases were ranked 1 through 4 (or 8 for high temperature parameters), where 4 (or 8) was best. The results are shown as radar plots in Figure 8.

The individual rankings for each grease were summed to give an overall score, shown next to the radar plots in Figure 8. Based on the overall score, for the testing parameters used here, the two lithium greases outperformed the two polyurea greases (SL = 34 > ML = 31 > SP = 25 > MP = 22). The overall score can also be separated into a friction rating and a wear rating, both of which are reported next to the radar plots in Figure 8. In terms of friction performance, the results indicate that the synthetic greases had better overall friction performance than the mineral based greases (SL = 19 > SP = 14 > ML = 10 > MP = 7). Consistent with the overall rating, SL exhibited the best friction performance, particularly at high temperatures. However, different trends are observed for wear, and the overall wear scores indicate that mineral greases outperformed their synthetic counterparts (ML = 21 > SL = 15, MP = 15 > SP = 11). The ML exhibited the best wear performance and, particularly, the least dependence of wear on surface roughness. MP and SL were tied for the second-best, but the good rating of SL was largely due to its low wear at high temperature whereas MP outperformed SL in all other wear metrics, particularly wear at low surface roughness.

Generally, the overall ratings, along with the radar charts themselves, serve as guidelines with which a designer can evaluate each grease based on metrics important for the application being considered. However, it should be noted that long duration and high speed bearing/grease life tests would be useful as another metric to include in this type of analysis for grease evaluation for EM applications.

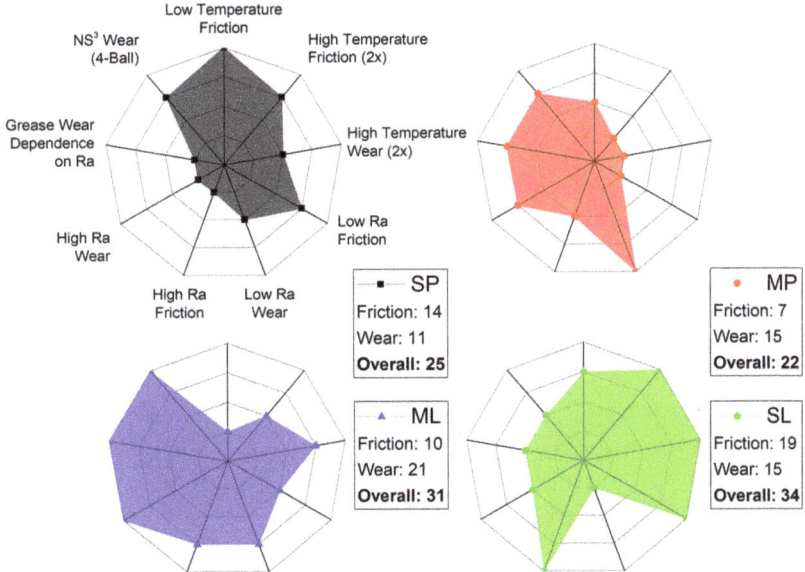

Figure 8. Grease ranking system based on a 1 to 4 scale (or 1 to 8 for high temperature parameters). A ranking of 4 (8) corresponds to best and 1 to worst. Low temperature is 40 °C, high temperature is 100–150 °C, low Ra is from 10–60 nm Ra, and high Ra from 120–200 nm Ra.

5. Conclusions

The tribological behavior of commercially available EM greases on hybrid bearing materials was characterized and evaluated. Results showed that EM grease products have notable differences in performance across different roughness and temperature conditions. These variations in performance have important implications for lubrication and design limitations. In general, greases whose performance is least affected by changing operating conditions will be more likely to meet the tribological needs of EMs.

Surface roughness has a significant impact on tribological properties. Rougher surfaces generally correspond to more friction and wear since they tend to have smaller local film thickness and higher pressure at asperity peaks [34,42,43], which result in high shear rates and stresses [12,42,43]. In contrast, surfaces with low roughness increase grease life and are less demanding in terms of lubrication ability since interacting smooth surfaces are more easily separated by lubricating films [8]. Yet, extremely smooth surfaces risk sudden seizure and asperities on rough surfaces may be useful for retaining lubricant [2]. Therefore, surface roughness plays a key role in determining the performance of grease lubricated systems.

Another important parameter for grease tribology is temperature, particularly for high-speed bearings. EM grease is known to be susceptible to thermo-oxidation degradation during high temperature bearing operation [9], which can lead to grease lubrication failure and, consequently, bearing failure [4,9–11]. Thus, grease formulations that do not compromise lubrication capabilities at high temperatures and resist thermal-oxidation are likely to be better for lubricating EM bearings. Temperature also influences film thickness through its effect on viscosity and the pressure-viscosity coefficient. Low temperature environments may be beneficial to achieving thicker lubricating grease films and reduce wear but will increase viscous friction. Further, excessively thick and viscous grease films may cause contact starvation from poor grease bleed and lack of reflow [16], which also leads to an increase in friction [22]. On the other hand, an increase in temperature can reduce viscous friction and activate grease bleed, but, under high temperatures, film thickness can decrease [16,22,44] to levels that may promote harsher operating conditions detrimental to grease and bearing life. Consequently, EM greases will need to be optimized for high operating temperatures and formulated to have minimal temperature dependence.

Both temperature and surface roughness affect the lubrication regime, as quantified by the λ ratio. Ideal λ ratios during operation will be small enough to achieve low friction but not so small that there is a transition into mixed or boundary lubrication. Ideal λ ratios can also have a positive effect on bearing contact fatigue, prolong component life, and improve energy efficiency. Therefore, maintaining a consistent λ ratio across temperature and roughness conditions is a key factor in component design and grease selection.

EM grease formulations also need to be optimized for hybrid bearing materials, assuming the continued use of hybrid bearings to combat stray current. Umbrella grease type products might not capture all lubrication requirements [34] and, consequently, may jeopardize performance and system life. Further, non-traditional bearing material and material configurations can exhibit wear mechanisms distinct from those observed in traditional steel bearings. The 4-ball test results reported here indicate the ideal hybrid bearing configuration is ceramic rolling elements on steel races (NS^3). The inverse bearing configuration, steel rolling elements on ceramic races (SN^3), generated significantly larger and abnormal wear. Additionally, the NS^3 configuration was found to have better wear performance than traditional SS^3 bearings, which has positive implications for hybrid bearings and grease life.

The results of a comprehensive set of friction and wear tests, using 4-ball tests and ball-on-disk measurements across a range of roughness and temperature conditions, showed that SL had the best overall performance under the conditions tested here (Figure 8). SL provided low wear at 40 nm Ra or less and consistently maintained low friction throughout both the full film and mixed lubrication regimes. When results were analyzed in terms of friction and wear separately, it was found that synthetic greases had the best friction behavior, while mineral greases had the best wear performance, with ML being best overall in terms of wear. However, ultimately grease selection will depend on the application. In the process of comparing four greases, this study also developed an approach for the λ ratio and the transition between lubrication regimes (Figure 7) that may be useful as a design tool more generally.

Going forward, the tribological performance of potential hybrid bearing materials combined with grease formulations for EMs need to be fully explored under conditions that resemble the environments of the target application. This is particularly important because tribology will play an important role enabling the electrification of the transportation industry, and, through tribological research, EM bearing lubrication can be optimized for EVs as it has been for ICEVs. In this context, the study reported here is a baseline and a template for further grease research in EM environments. Further, the present study demonstrates that market-available EM grease products can vary significantly in performance, providing insight into the effects of operating conditions and design criteria on grease behavior.

Author Contributions: Conceptualization, A.M. and D.S.G.; methodology, A.M. and D.S.G.; experiments, S.L. and D.S.G.; data curation, A.M. and D.S.G.; writing—original draft preparation, A.M. and D.S.G.; writing—review and editing, A.M., S.L. and D.S.G. All authors have read and agreed to the published version of the manuscript.

Funding: This research was supported by the National Lubricating Grease Institute (NLGI).

Acknowledgments: We appreciate the valuable input from our NLGI liaison throughout the project. We also acknowledge the UC Merced Instructional Lab Support Team and Instrumentation Foundry. Lastly, some of the 4-ball tests were performed by undergraduate students in the research group, Colin Cox, Alex McCollum, Jose Morales, and Eddie Santiago.

Conflicts of Interest: The authors declare no conflict of interest.

References

1. Nieuwenhuis, P.; Cipcigan, L.; Sonder, H.B. The Electric Vehicle Revolution. In *Future Energy*; Elsevier: Amsterdam, The Netherlands, 2020; pp. 227–243.
2. Stachowiak, G.; Batchelor, A. *Engineering Tribology*; Butterworth-Heinemann: Oxford, UK, 2013.
3. Farfan-Cabrera, L.I. Tribology of electric vehicles: A review of critical components, current state and future improvement trends. *Tribol. Int.* **2019**, *138*, 473–486. [CrossRef]
4. He, F.; Xie, G.; Luo, J. Electrical bearing failures in electric vehicles. *Friction* **2020**, *8*, 4–28. [CrossRef]
5. Lukaszczyk, M. Improving efficiency in electric motors. *World Pumps* **2014**, *2014*, 34–41. [CrossRef]
6. de Santiago, J.; Bernhoff, H.; Ekergård, B.; Eriksson, S.; Ferhatovic, S.; Waters, R.; Leijon, M. Electrical Motor Drivelines in Commercial All-Electric Vehicles: A Review. *IEEE Trans. Veh. Technol.* **2012**, *61*, 475–484. [CrossRef]
7. Walther, H.C.; Holub, R.A. Lubrication of electric motors as defined by IEEE standard 841-2009, shortcomings and potential improvement opportunities. In Proceedings of the 2014 IEEE Petroleum and Chemical Industry Technical Conference (PCIC), San Francisco, CA, USA, 8–10 September 2014; pp. 91–98.
8. Lugt, P.M. Modern advancements in lubricating grease technology. *Tribol. Int.* **2016**, *97*, 467–477. [CrossRef]
9. Yu, Z.; Yang, Z. Fatigue failure analysis of a grease-lubricated roller bearing from an electric motor. *J. Fail. Anal. Prev.* **2011**, *11*, 158–166. [CrossRef]
10. Fernandes, P. Contact fatigue in rolling-element bearings. *Eng. Fail. Anal.* **1997**, *4*, 155–160. [CrossRef]
11. Fernandes, P.; McDuling, C. Surface contact fatigue failures in gears. *Eng. Fail. Anal.* **1997**, *4*, 99–107. [CrossRef]
12. Kanazawa, Y.; Sayles, R.S.; Kadiric, A. Film formation and friction in grease lubricated rolling-sliding non-conformal contacts. *Tribol. Int.* **2017**, *109*, 505–518. [CrossRef]
13. Cann, P. Grease lubrication of rolling element bearings—Role of the grease thickener. *Lubr. Sci.* **2007**, *19*, 183–196. [CrossRef]
14. Cousseau, T.; Graça, B.; Campos, A.; Seabra, J. Grease aging effects on film formation under fully-flooded and starved lubrication. *Lubricants* **2015**, *3*, 197–221. [CrossRef]
15. Zapletal, T.; Sperka, P.; Krupka, I.; Hartl, M. On the Relation between Friction Increase and Grease Thickener Entraining on a Border of Mixed EHL Lubrication. *Lubricants* **2020**, *8*, 12. [CrossRef]
16. Cann, P. Starved grease lubrication of rolling contacts. *Tribol. Trans.* **1999**, *42*, 867–873. [CrossRef]
17. De Laurentis, N.; Kadiric, A.; Lugt, P.; Cann, P. The influence of bearing grease composition on friction in rolling/sliding concentrated contacts. *Tribol. Int.* **2016**, *94*, 624–632. [CrossRef]
18. Cen, H.; Lugt, P.M.; Morales-Espejel, G. On the film thickness of grease-lubricated contacts at low speeds. *Tribol. Trans.* **2014**, *57*, 668–678. [CrossRef]
19. Cen, H.; Lugt, P.M. Film thickness in a grease lubricated ball bearing. *Tribol. Int.* **2019**, *134*, 26–35. [CrossRef]
20. Cen, H.; Lugt, P.M.; Morales-Espejel, G. Film thickness of mechanically worked lubricating grease at very low speeds. *Tribol. Trans.* **2014**, *57*, 1066–1071. [CrossRef]
21. Morales-Espejel, G.; Lugt, P.M.; Pasaribu, H.; Cen, H. Film thickness in grease lubricated slow rotating rolling bearings. *Tribol. Int.* **2014**, *74*, 7–19. [CrossRef]
22. Gonçalves, D.E.; Campos, A.V.; Seabra, J.H. An experimental study on starved grease lubricated contacts. *Lubricants* **2018**, *6*, 82. [CrossRef]
23. Sebastian, T. Temperature effects on torque production and efficiency of PM motors using NdFeB magnets. *IEEE Trans. Ind. Appl.* **1995**, *31*, 353–357. [CrossRef]
24. Hamidizadeh, S.; Alatawneh, N.; Chromik, R.R.; Lowther, D.A. Comparison of different demagnetization models of permanent magnet in machines for electric vehicle application. *IEEE Trans. Magn.* **2016**, *52*, 1–4. [CrossRef]
25. Lugt, P.M. *Grease Lubrication in Rolling Bearings*; John Wiley & Sons: Hoboken, NJ, USA, 2012.
26. Oliver, J.; Guerrero, G.; Goldman, J. Ceramic bearings for electric motors. In Proceedings of the 2015 IEEE-IAS/PCA Cement Industry Conference (IAS/PCA CIC), Toronto, ON, Canada, 26–30 April 2015; pp. 1–11.
27. Gonda, A.; Capan, R.; Bechev, D.; Sauer, B. The Influence of Lubricant Conductivity on Bearing Currents in the Case of Rolling Bearing Greases. *Lubricants* **2019**, *7*, 108. [CrossRef]
28. Dante, R.C.; Kajdas, C. A review and a fundamental theory of silicon nitride tribochemistry. *Wear* **2012**, *288*, 27–38. [CrossRef]
29. Volante, M.; Fubini, B.; Giamello, E.; Bolis, V. Reactivity induced by grinding in silicon nitride. *J. Mater. Sci. Lett.* **1989**, *8*, 1076–1078. [CrossRef]
30. Zaretsky, E.V.; Vlcek, B.L.; Hendricks, R.C. Effect of silicon nitride balls and rollers on rolling bearing life. *Tribol. Trans.* **2005**, *48*, 425–435. [CrossRef]
31. Johnson, D.W. *The Tribology and Chemistry of Phosphorus Containing Lubricant Additives*; IntechOpen: London, UK, 2016.
32. Bertrand, P. Reactions of tricresyl phosphate with bearingmaterials. *Tribol. Lett.* **1997**, *3*, 367–377. [CrossRef]
33. Guan, B.; Pochopien, B.A.; Wright, D.S. The chemistry, mechanism and function of tricresyl phosphate (TCP) as an anti-wear lubricant additive. *Lubr. Sci.* **2016**, *28*, 257–265. [CrossRef]
34. Gonçalves, D.; Vieira, A.; Carneiro, A.; Campos, A.V.; Seabra, J.H. Film thickness and friction relationship in grease lubricated rough contacts. *Lubricants* **2017**, *5*, 34. [CrossRef]
35. Chang, H.; Lan, C.; Chen, C.; Kao, M.; Guo, J. Anti-wear and friction properties of nanoparticles as additives in the lithium grease. *Int. J. Precis. Eng. Man.* **2014**, *15*, 2059–2063. [CrossRef]

36. Qiang, H.; Anling, L.; Yangming, Z.; Liu, S.; Yachen, G. Experimental study of tribological properties of lithium-based grease with Cu nanoparticle additive. *Tribol.-Mater. Surf. Interfaces* **2017**, *11*, 75–82. [CrossRef]
37. Qiang, H.; Anling, L.; Yachen, G.; Songfeng, L.; LH, K. Effect of nanometer silicon dioxide on the frictional behavior of lubricating grease. *Nanomater. Nanotechnol.* **2017**, *7*, 1–9.
38. Gow, G. Lubricating grease. In *Chemistry and Technology of Lubricants*; Springer: Berlin/Heidelberg, Germany, 1992; pp. 255–268.
39. Strunks, G.; Toth, D.; Saba, C. Geometry of wear in the sliding four-ball wear test. *Tribol. Trans.* **1992**, *35*, 715–723. [CrossRef]
40. Walters, N.; Martini, A. Friction dependence on surface roughness for castor oil lubricated NiTi alloy sliding on steel. *Tribol. Trans.* **2018**, *61*, 1162–1166. [CrossRef]
41. Hansen, J.; Björling, M.; Larsson, R. Lubricant film formation in rough surface non-conformal conjunctions subjected to GPa pressures and high slide-to-roll ratios. *Sci. Rep.* **2020**, *10*, 22650. [CrossRef]
42. Guegan, J.; Kadiric, A.; Spikes, H. A study of the lubrication of EHL point contact in the presence of longitudinal roughness. *Tribol. Lett.* **2015**, *59*, 22. [CrossRef]
43. Guegan, J.; Kadiric, A.; Gabelli, A.; Spikes, H. The relationship between friction and film thickness in EHD point contacts in the presence of longitudinal roughness. *Tribol. Lett.* **2016**, *64*, 33. [CrossRef]
44. Cann, P. Starvation and reflow in a grease-lubricated elastohydrodynamic contact. *Tribol. Trans.* **1996**, *39*, 698–704. [CrossRef]

 *lubricants*

Article
Testing Grease Consistency

Alan Gurt and Michael M. Khonsari *

Department of Mechanical and Industrial Engineering, Louisiana State University, Baton Rouge, LA 70803, USA; agurt2@lsu.edu
* Correspondence: khonsari@lsu.edu

Abstract: Because of the influential role of consistency in selecting a grease for a given application, accurate and meaningful methodologies for its measurements are vitally important. A new method, recently introduced, uses a rheometer to compress a grease sample to evaluate a relative consistency between a fresh and degraded grease; however, the results of this approach compared to a standard penetrometer and other methods of assessing consistency have not been studied. This paper takes a closer look at the relevant parameters involved in the rheometer penetration test and establishes a recommended procedure for its use. The consistency of various greases is then tested using this method and compared to results obtained from yield stress, crossover stress, and cone penetration tests. The results indicate that rheometer penetration may be used to assess the change in consistency for a given grease but should not be used to compare different greases. For this purpose, the crossover stress method is recommended, which is shown to correlate very well with cone penetration while using a simple procedure and allowing the use of a substantially smaller sample. A strong power law correlation between crossover stress and cone penetration was found for all greases tested and is presented in Figure 12.

Keywords: grease consistency; grease testing; cone penetration; rheometer testing

1. Introduction

The consistency of a lubricating grease is often considered to be its most important rheological property. It generally dictates the suitability of a grease for a particular application. Consistency broadly refers to the firmness of a grease, indicating a grease's ability to remain in place (resist leakage) within bearings and to form stable channels of lubricant [1]. These channels are important to the functionality of a bearing because they serve as reservoirs from which moving parts draw lubricating fluid over the operational lifetime of a machine [2]. Hence, consistency serves as an important metric for selecting a grease, and the capability to properly quantify this property is of great importance.

The main test to measure grease consistency is the cone penetration test given by ASTM D217 [3]. In this test, a cone is dropped into a grease sample for 5 s and the depth to which the cone penetrates is used as a measure of consistency. This test requires a large sample of grease—approximately 450 grams according to ASTM D217—so the alternative scaled-down tests given by ASTM D1403 [4] are able to provide penetration results with smaller samples. The penetration depth obtained from any of these tests is usually used to assign a grade to a grease for succinctly characterizing its nature to a broad market of consumers.

Using specialized equipment such as a rheometer is becoming increasingly popular for assessing the rheological properties of grease. This apparatus requires a very small sample—less than two grams—and provides quantitative results that can assess grease consistency. It is perhaps even more precise than the cone test with the advantage of requiring a simpler operational procedure while allowing for temperature control. These tests are often used to evaluate a "critical" stress but have also been used as a way of conducting a penetration test [5,6]. The methodology of conducting a penetration test is quite simple, whereas the calculation of critical stresses demands an understanding of viscoelasticity.

Citation: Gurt, A.; Khonsari, M.M. Testing Grease Consistency. *Lubricants* **2021**, *9*, 14. https://doi.org/10.3390/lubricants9020014

Received: 5 January 2021
Accepted: 30 January 2021
Published: 2 February 2021

Publisher's Note: MDPI stays neutral with regard to jurisdictional claims in published maps and institutional affiliations.

Copyright: © 2021 by the authors. Licensee MDPI, Basel, Switzerland. This article is an open access article distributed under the terms and conditions of the Creative Commons Attribution (CC BY) license (https://creativecommons.org/licenses/by/4.0/).

A viscoelastic material is one that has properties of both a liquid and a solid. Grease, for instance, can be considered a viscoelastic solid material since it behaves as a solid unless sheared. Once sheared past a critical point, it begins to flow. The quantification of this state is of great interest for characterizing a substance, but the exact definition of this state has been a subject of debate for years [7]. In addition to having different definitions, there are various methods for evaluating this state, including the use of steady flow curves, creep measurements, stress ramps, stress sweeps, and oscillatory tests [8,9]. Nevertheless, in the context of lubricating grease, two distinct points found through oscillatory tests—the yield point and flow point—have come to be popular choices for characterizing a grease.

Oscillatory rheometry involves the analysis of viscoelastic materials by monitoring their stress response to oscillating strain (or vice-versa). The solid-like behavior is characterized by the storage modulus, which describes the ability of a material to store energy through elastic deformation and is in phase with the oscillating input. Alternatively, the liquid-like behavior is characterized by the loss modulus, which describes the ability of a material to lose energy through viscous dissipation and is out of phase with the input. These two parameters both change as the state of stress within material changes, so imposing an increasing stress or strain amplitude (often called an amplitude sweep) in an oscillatory test is often an ideal way of examining viscoelastic behavior. Results of an amplitude sweep can then be used to calculate the yield point and/or flow point.

The yield point, characterized by the yield stress, is considered by many to be the end of the linear viscoelastic range at which the structure begins to be considered damaged. One choice is to define the linear viscoelastic range as the region where the storage modulus is independent of strain. Another choice is to use the end of stress-strain linearity [10] as the yield stress. Due to the recent popularity of the second choice in analyzing grease [9–14], it will be used as the definition of yield stress throughout this paper.

The flow point, characterized by the crossover stress, is the point at which the storage modulus and loss modulus cross over each other. Upon commencing an amplitude sweep, the storage modulus will quickly reach a plateau value, but eventually start decreasing. The flow point will be considered the point where the storage modulus decreases enough to cross over the loss modulus, temporarily signifying a material with properties more closely resembling a liquid than a solid. Both this point and the yield point are indications of how "solid" a given material is and are considered properties akin to consistency.

Over time, overall grease consistency can permanently change due to the shear irreversibly breaking the structure, due to oil bleed, leakage, and evaporation, due to contamination, and due to chemical reactions [15]. This change in consistency has been used to track the degree of degradation a grease has undergone [5,16,17]. A sampling of in-service grease in conjunction with a degradation model can be used to estimate the remaining useful life. However, the cone penetration tests given by ASTM D217 and ASTM D1403 require a larger sample of grease than is used for many applications. This means the consistency of an in-service grease cannot be measured with these methods. Therefore, it is advantageous to have alternative consistency tests that allow the use of a small sample of grease. One such test intended for field use is provided in a test kit given by the bearing manufacturer SKF, but the resolution of the results is limited to the grade. Therefore, rheometer tests allowing more precise results and temperature control using a very small sample have a significant value. A closer examination of these tests is the main focus of this paper.

This paper examines common consistency tests in Section 2, takes a closer look at a proposed method for assessing consistency through a rheometer penetration test in Section 3, compares results of these tests to each other in Section 4, presents a discussion of results in Section 5, and gives concluding remarks in Section 6.

2. Examination of Consistency Tests

The most common tests of grease consistency will be overviewed and a discussion of each is provided.

2.1. Cone Penetration Test

The accepted method for testing grease consistency is the cone penetration test given by ASTM D217 [3]. In this test, a cone is dropped into a cup of grease and the depth of penetration is used as a measurement of consistency. This value of penetration is used to assign a grade to a grease from 000 to 6. The National Lubricating Grease Institute (NLGI) defines the ranges of penetration corresponding to the grade as given in Table 1. The values of penetration are given in tenths of a millimeter, often called decimillimeters and abbreviated as dmm.

Table 1. National Lubricating Grease Institute (NLGI) grades and applications [18].

NLGI Grade	Penetration [dmm]	Food Equivalent	Common Application
000	445–475	Ketchup	Gear boxes/low temperature use
00	400–430	Yogurt	Gear boxes/low temperature use
0	355–385	Mustard	Centralized lubrication systems
1	310–340	Tomato paste	General purpose bearings
2	265–295	Peanut butter	General purpose bearings
3	220–250	Butter	High-speed bearings
4	175–205	Frozen yogurt	Very high-speed bearings
5	130–160	Fudge	Low-speed journal bearings
6	85–115	Cheddar cheese	Very slow journal bearings

The grades in this scale are defined so that they span 30 dmm and have gaps of 15 dmm between grades. This leads to some ambiguity in labeling a grease, where many choose to assign half grades. For instance, some would choose to assign a grade of 2.5 to a grease with a worked penetration of 260 dmm.

As the cone penetrates deeper into the grease sample, grease begins to lift out of the cup as is demonstrated in Figure 1. As penetration increases, this geometry-dependent behavior begins to influence penetration readings. ASTM D217 indicates that the penetration for soft greases is a function of cup diameter if its penetration is above 265 dmm. However, the grease eventually reaches a point where the complicated nature of the squeezing of grease between the cone and the lip of the cup becomes even more of a determinant of the final penetration value. To account for this, the standard mandates that the cone be perfectly centered when the penetration reading is above 400 dmm.

Not only does the cone penetration test require a substantial quantity of grease, it also demands a skilled operator in order to obtain consistent results. A significant source of error in this test is the presence of pockets of air within the grease sample. This is particularly relevant for measuring the unworked penetration of a grease sample, as loading the sample into the grease cup can easily form large pockets of air. Therefore, having an unskilled operator can cause inconsistent measurements of consistency using the cone penetration method.

Figure 1. Grease cup (**a**) before and (**b**) after cone penetration.

2.2. Rheometer Oscillatory Tests

A rheometer—such as the one pictured in Figure 2—has the capacity to perform numerous useful tests to analyze viscoelastic materials. Through oscillatory strain sweep measurements, properties of viscoelastic materials can be found without potential errors from sample fracturing [19] and with minimal sensitivity to gap height and plate roughness [9,10,20]. Two parameters will be investigated here: the yield point—corresponding with the yield stress—and the flow point—corresponding with the crossover stress. The definition set forth by Cyriac et al. [10] will be used for yield stress, where the end of stress-strain linearity measured from an oscillatory amplitude sweep is considered the yield point. For the crossover stress, the first point at which the storage modulus and loss modulus reach the same value will be considered the definition.

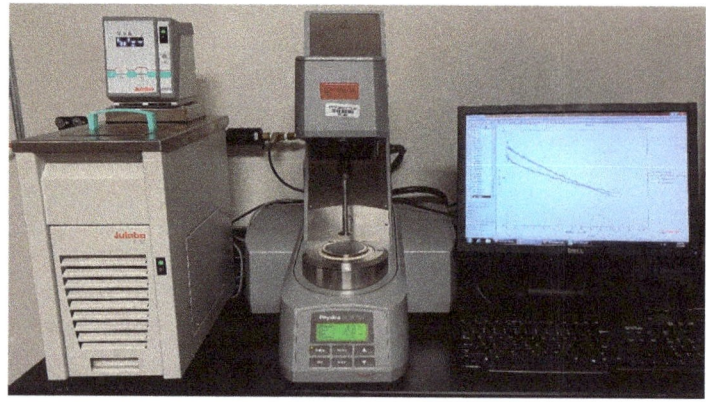

Figure 2. Anton Paar MCR 301 rheometer used in experiments

When performing these tests, there are many choices for geometry, but parallel-plate geometry is perhaps the most common. In addition, there are many choices of plate diameters, but 25 mm and 50 mm are among the most common. For rheological measurements, a smaller diameter will measure higher stresses than a larger diameter if all other variables are held constant. Therefore, the results obtained from different plate diameters cannot be reliably compared. Nevertheless, a smaller diameter is generally advantageous because it is easier to load the sample, and only a small quantity of the sample is needed despite having slightly reduced testing precision.

Another important consideration for rheometer tests is the sample loading procedure. In fact, different methods of loading a sample can cause errors of up to 30% in some cases [21]. The typical procedure is to apply a sample to the lower and/or upper plate and then lower the top plate to a gap slightly larger than the measurement gap. At this point, a tool is used to clear (trim) all the sample that is not directly between the plates. The top plate is then lowered to the measurement gap and the measurement is performed. A standard procedure is to trim the sample at a gap 5% greater than the measurement gap, but trimming the sample at 2.5% above the measurement gap is also reasonable.

Another consideration is the state of stress within the sample as it is being measured. Upon lowering the top plate, there can be a large resistive force acting on the top plate due to the compression of the sample. This effect is especially pronounced when using a plate with a large diameter. This is generally dealt with in one of three ways: waiting for the sample to relax (relaxation), shearing the sample a small amount (pre-shear), or doing nothing and immediately commencing measurement. For many measurements, this choice is inconsequential. However, in some cases, this can have a significant effect, and this is explored in Sections 3 and 4.

Overall, performing these measurements requires an operator to learn how to use a rheometer, but the actual treatment of a sample is quite straightforward and far less prone to operator error than the cone penetration test.

2.3. Alternative Penetration Tests

Though the cone penetration tests given by ASTM D217 and ASTM D1403 are the officially recognized penetration tests, there are other penetration tests that can possibly be used to assess grease consistency. Two of the most prominent include the consistency test found within the SKF grease test kit [22] and the rheometer penetration test.

The consistency test provided in the SKF grease test kit is intended for the in-service sampling of grease and needs only a very small sample. This consistency test is an example of a constant-volume squeeze flow [23]. It is done by first applying a cylindrical-shaped sample of grease to a glass plate using a jig. Then, another glass plate is carefully placed above this one and a weight is put atop it. The weight is allowed to cause the grease to spread between the plates for 15 s and the final diameter of the sample is compared to rings on a sheet of paper to determine consistency. This only has the resolution to determine the NLGI grade and relies on the final sample shape being close to circular. In practice, a used grease sample may be quite nonhomogeneous and contaminated, leading to a non-circular spread, which will add difficulty to determining the result.

The rheometer penetration test has been used in the past [5,6] but will be examined more in-depth in Section 3.

3. Details of Rheometer Penetration Test

The geometry given by the rheometer penetration test is an example of "imperfect squeeze flow" [23,24], with the geometry having neither a constant area nor constant volume. In this configuration, there is a complicated variable pressure imposed by the fluid squeezed as the gap closes [23]. This leads to some difficulty in deriving analytical equations describing the relationships among parameters, such as displacement, force, area, and velocity. Therefore, these parameters were investigated empirically.

Many variables were identified for the rheometer penetration test, including sample preparation, gap height, penetration time, and normal force imposed. Similar to the oscillatory tests done in a rheometer, the test geometry and sample preparation are key variables that must be arbitrarily chosen. For the same reasons as in the oscillatory tests, it appears that the 25 mm diameter plate configuration is a good choice, and this was used in the previous works mentioned.

All rheometer experiments were conducted using an Anton Paar MCR 301 rheometer at room temperature. Each sample was measured three times and the average value is presented with error bars corresponding to one standard deviation. In most cases, the standard deviation is quite low and error bars are not visible. An overview of the greases used in the experiments is provided in Table 2.

Table 2. Greases used.

Grease Abbreviation	Thickener Type	Labeled NLGI Grade
LiC00	Lithium complex	00
LiC0	Lithium complex	0
LiC1	Lithium complex	1
LiC2	Lithium complex	2
LiC3	Lithium complex	3
AlC2.1	Aluminum complex	2 *
AlC2.2	Aluminum complex	2
CaS2	Calcium sulfonate	2
PU2	Polyurea	2

* The measured consistency of this grease was approximately grade 1.

In choosing the gap height, the main constraint is that the ratio of the plate's radius to gap height should be greater than or equal to 10 to avoid edge effects [25]. The selection of the plate diameter is another important parameter, as results of similar experimentation [24] indicate that there is a complicated dependency of rheological measurements on plate geometry. For these measurements, a 25 mm diameter top plate was selected. This is a common size and can allow for using a small sample of grease while allowing a reasonable gap. In order to satisfy the radius-gap constraint given above for a 25 mm diameter, the gap must be below 1.25 mm. Hence, a 1 mm gap was chosen for the experiments reported in this paper.

The next thing to be examined is sample preparation. Measurements of the same sample were taken using a standard 5% trim with no pre-shear, a 100% trim without pre-shear, a 5% trim with pre-shear, and a 5% trim with a 20-min relaxation period. Using a 5% trim without pre-shear or a relaxation period means that the top plate is initially lowered to a height 5% above the measurement gap where the sample is trimmed before lowering the top plate to the measurement gap and immediately commencing the measurement. Using a 100% trim means the top plate is initially lowered to twice the measurement gap where the sample is trimmed before lowering the plate and immediately commencing measurement. The 5% trim with pre-shear means the sample was trimmed at 5% above the measurement gap but after the plate was lowered to the measurement gap, a shear rate of 5 s^{-1} was induced for 5 s before commencing measurement. Finally, the 5% trim with relaxation period means that after trimming the sample at 5% above the measurement gap, the top plate is lowered to the measurement gap and a pause of 20 min is taken before commencing measurement.

Results of the sample preparation investigation are displayed in Figure 3 and show that using pre-shear does not appear to influence the result when compared to the 5% trim. Allowing the sample to relax for 20 min before subjecting it to penetration appeared to slightly decrease the penetration, but also led to a higher standard deviation in this case. Adding 20 min to a 20 s test is also impractical. Nevertheless, the method with the lowest standard deviation of results is the case of the overfilled gap, where the sample was trimmed at 2 mm and immediately subjected to penetration. This method yielded a consistently lower penetration when compared to the 5% trim, which is an expected result when considering the nature of squeezing flow. The repeatability of this method led to its implementation in subsequent measurements.

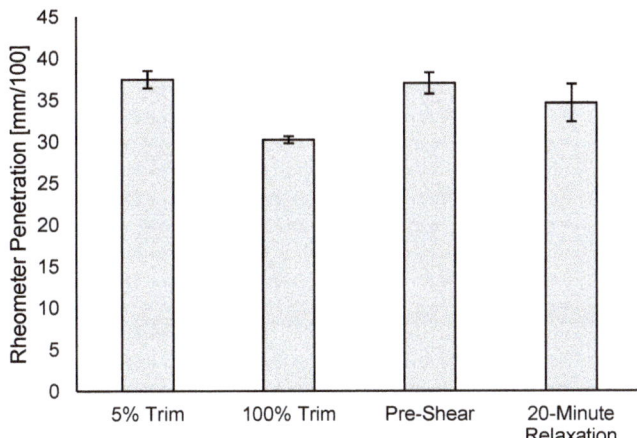

Figure 3. Rheometer penetration results of PU2 with a 4 N normal force comparing a standard 5% trim to a 100% trim, a pre-sheared sample, and to a sample relaxed for 20 min.

To select an appropriate penetration time, one must examine the penetration as a function of time. Figure 4 examines the gap height over time as a constant normal force is imposed on grease samples. This figure considers a range of forces between 2 N and 8 N and uses greases ranging from grade 00 to grade 2. For some grease types, a low normal force will cause the plate to barely move or, in some cases, not move at all. In cases where the top plate moves, the initial velocity (slope of this plot) is approximately the same but begins to level off as the plate nears a steady state. Because of this, it takes longer for the top plate to reach a steady state when the plate must travel a longer distance. Nevertheless, for all cases, it appears that 20 s is sufficient, confirming the results of a previous study [6]. Therefore, 20 s was used for all subsequent measurements.

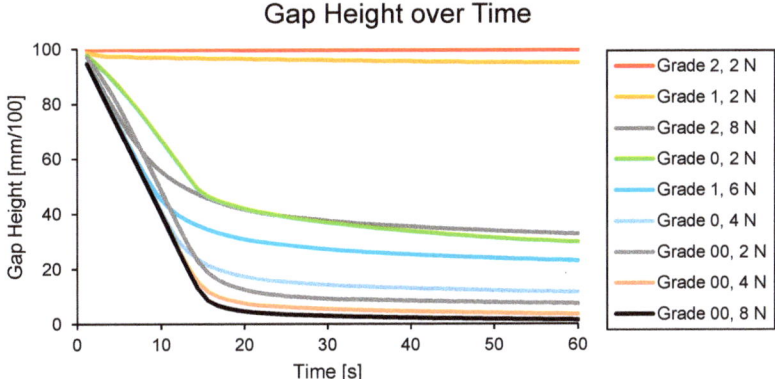

Figure 4. Rheometer gap height as a function of time for various greases and forces.

The next thing to look at is how penetration is affected by changing the normal force. For this examination and all future rheometer penetration test results, data will be presented as net penetration to provide a similar meaning as cone penetration. This will be defined as the final gap height subtracted from the initial gap height. In this way, the higher the penetration, the less firm a given grease is. In addition, the units of mm are converted to mm/100, which also corresponds to the percentage of the initial gap.

Figure 5 shows that there is a clearly nonlinear trend relating net penetration and normal force for a given grease sample. It is possible to separate the trend into three regions: a region where the force is too low, a region where the force is too high, and a region in the middle where the force is appropriate for measuring consistency. For greases below grade 1, the first region is not visible on the plot because imposing a force of less than 1 N using the rheometer often led to no movement. A force of 1.2 N would work well to determine the consistency of a grade 00 or grade 0 grease. However, it would not be able to determine any significant difference between grade 1 and grade 2 grease. Similarly, a force of 4 N would work well for a grade 1 or grade 2 grease but would not be useful for any grease below grade 1. Generally, the goal of testing greases with the rheometer penetration method is to keep results within the middle region. This means that the rheometer penetration test needs to vary the force according to the consistency of the grease chosen in order to keep results within the region of approximately between 15 mm/100 and 80 mm/100. Unfortunately, this means that if the rheometer penetration results are to be correlated with other tests, each normal force must be considered individually.

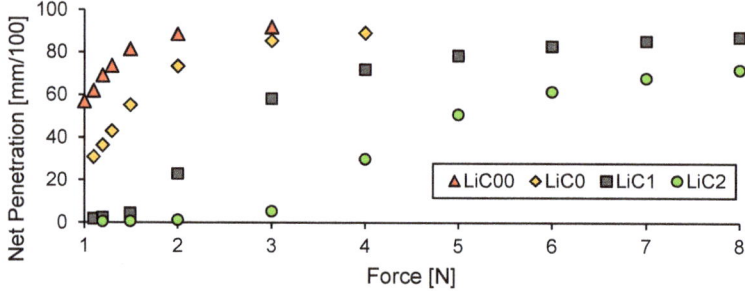

Figure 5. Rheometer net penetration as a function of force for various grease grades.

The final procedure developed for assessing the consistency of a grease using the rheometer penetration test involves first applying the sample to the base plate and/or upper plate of the rheometer. The top plate is then lowered to 2 mm where the sample is trimmed, and then lowered to the measurement gap of 1 mm. The desired normal force is imposed for 20 s, and the difference between the initial and final value is reported as the net penetration value. If the net penetration is not between 15 mm/100 and 80 mm/100, then the force should be changed.

4. Comparison of Tests

The procedure developed for testing grease consistency using a rheometer penetration test will now be compared with the cone penetration test, yield stress measurements, and crossover stress measurements to observe correlations among these tests at room temperature.

4.1. Materials and Procedures

In order to test as wide a range of consistencies as possible and to obtain various intermediate consistencies, greases (given in Table 2) were sheared for various intervals in a standard grease worker specified by ASTM D217. For many of the greases, such as PU2 and AlC2.1, this led to considerable changes in consistency. However, some of the other grease types, such as all the lithium complex greases tested, showed a minimal change in consistency from the mechanical shear of the grease worker. In order to obtain an even wider array of consistencies, the lithium complex greases were mixed together in various ratios and the calcium sulfonate grease was contaminated with small amounts of water (under 10 percent by weight). The procedure for mixing involves placing the desired ratio of materials together in the grease worker and working for a minimum of 500 strokes. Overall, the exact proportions of greases, the exact concentration of water, and the exact number of strokes are not relevant to this investigation. This study is exclusively focused on examining the similarities and differences among the various consistency tests using the same sample in each.

Unless indicated otherwise, all results of oscillatory tests (calculation of yield stress and crossover stress) were conducted using a 25 mm flat plate with a 1 mm gap at 1 Hz with no relaxation or pre-shear. These tests, as well as cone penetration measurements and rheometer penetration measurements, were done three times, and the average value is presented.

4.2. Identifying Variables

A further investigation of variables was conducted for the oscillatory tests. These tests calculated the yield stress and crossover stress of one particular sample but assessed the influence of plate size, plate roughness, overfilling the gap, and pre-shearing the sample. The results are summarized in Figure 6.

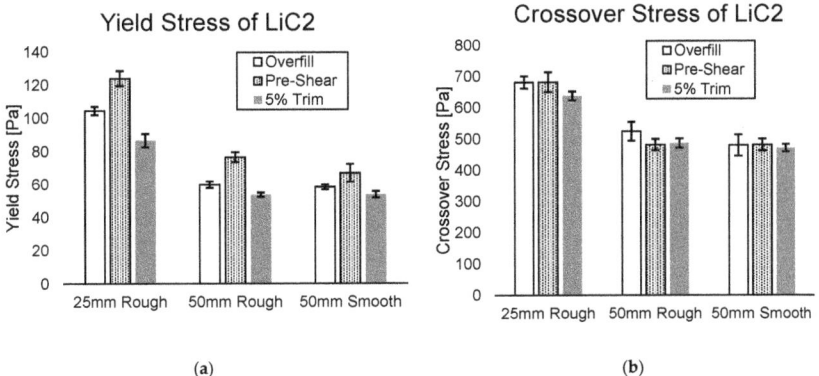

(a) (b)

Figure 6. Comparison of (**a**) yield stress and (**b**) crossover stress of LiC2 worked for 60 strokes using an overfilled sample compared to a pre-sheared sample compared to a standard 5% trim.

The results based on plate properties indicate that, as expected, the smaller plate measured a higher yield stress and crossover stress than the larger plates. In addition, it appears that the surface roughness of the plate does not cause a significant difference in the results. This is also expected, as one of the advantages of oscillatory rheometry is the minimal sensitivity to plate roughness.

A pre-shear of $5\ s^{-1}$ for 5 s prior to measuring the critical stresses was done to intensify the effects of sample manipulation during testing and assess the sensitivity of the test to initial conditions. This investigation revealed that pre-shear had a significant effect on calculating yield stress but had a small effect on calculating crossover stress. Due to the relatively large standard deviation observed when using pre-shear, it was not used for subsequent tests.

Overfilling of the sample with a 100% trim was found to significantly affect results for both yield stress and crossover stress. In addition to overestimating both stresses, a higher standard deviation was noticed for results obtained using an overfilled gap. Therefore, care was taken to cleanly trim every sample at 5% above the measurement gap.

4.3. Rheometer Penetration Test Results

Here, the results of all four tests considered (cone penetration, rheometer penetration, yield stress, and crossover stress) will be compared. The first set of results presented involves looking more closely at the rheometer penetration test. Because different normal forces must be used to test different greases, here, each normal force has an associated set of data. For simplicity, only one set of data will be presented; however, this set is particularly representative of all data collected. Figures 7 and 8 compare the rheometer penetration test done using a 5 N normal force to other consistency tests.

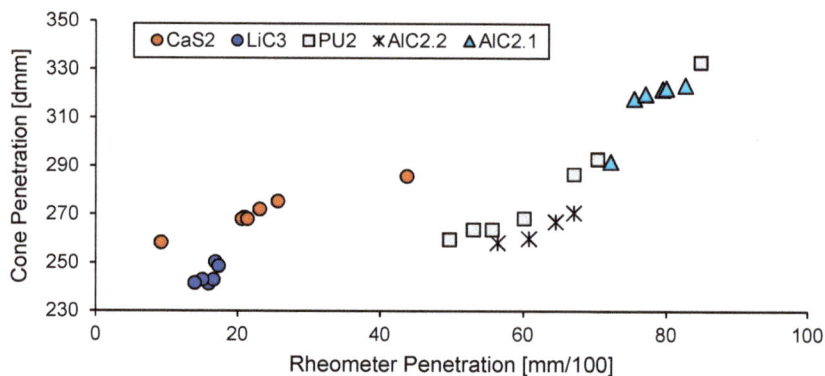

Figure 7. Cone penetration vs rheometer penetration for various greases with a 5 N normal force.

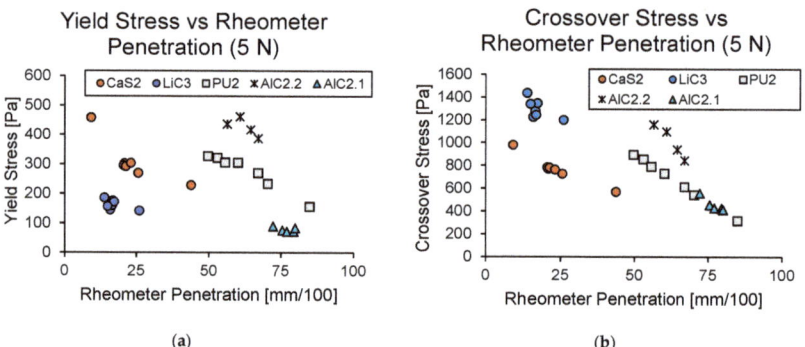

Figure 8. Comparison of (**a**) yield stress and (**b**) crossover stress with rheometer penetration at a force of 5 N for various grease types.

Figure 7 shows that there is a general positive correlation between cone penetration and rheometer penetration; however, different greases show slightly different behaviors. This means that using a rheometer to assess the change to a particular grease's consistency would be a valid approach, but comparing two different greases to each other with this approach would not. Using a linear correlation between cone penetration and rheometer penetration in this case would likely lead to excessive error and it is, therefore, recommended that this approach not be taken if one is interested in comparing the consistency of different greases.

Results shown in Figure 8 are perhaps even more indicative that a correlation between rheometer penetration and the other tests that should not be used in practice since a meaningful trend cannot be established. Nevertheless, a comparison of the other tests yielded interesting results.

4.4. Oscillatory Test Results

The next results shown compare the oscillatory test results to each other and to cone penetration results.

Figure 9 shows the relationship between yield stress and crossover stress for the greases selected. As expected, based on another study [8], there is a general positive correlation between these two measurements. However, there is one particularly notable cluster of data points associated with the LiC3 grease. These points were found to completely deviate from the expected trend and indicate that there is some important discrepancy between yield stress and crossover stress.

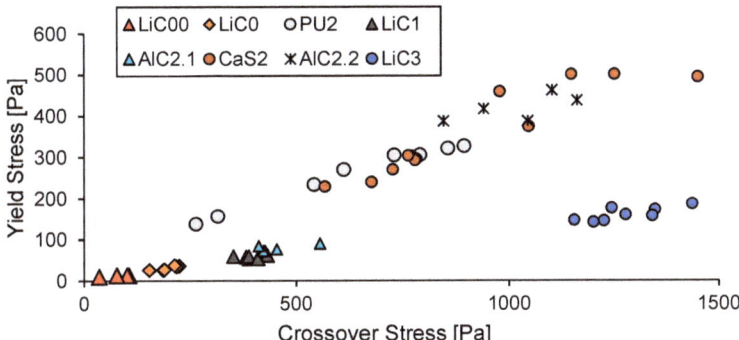

Figure 9. Yield stress vs crossover stress for various greases.

Figure 10 shows the relationship between cone penetration and yield stress, where a general negative correlation is established. This figure, once again, shows that the LiC3 grease deviates from the expected trend by a significant amount. This grease appears to have a much lower cone penetration value than the PU2 grease, yet it shows a similar value of yield stress. Because of this, it is not recommended to use this measure of yield stress to estimate cone penetration.

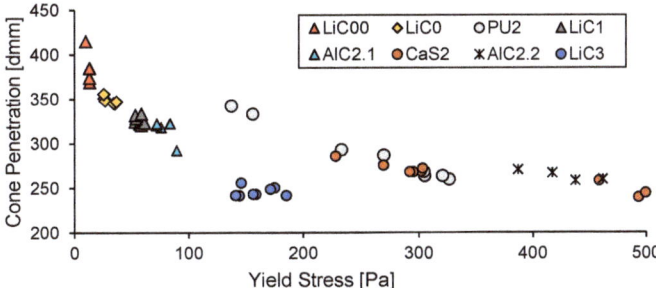

Figure 10. Cone penetration vs yield stress for various greases.

Figure 11 shows a plot of cone penetration (CP) vs. crossover stress (CS), where the correlation is significantly better than in any of the preceding plots. In this case, the LiC3 grease fits in exactly where it would be expected and no grease deviates significantly from the overall trend. These data are fitted to a power law in Figure 12, where the coefficient of determination is above 0.95. Because of the good fit for all types of greases tested, it appears that using a rheometer to measure crossover stress could be used as a reasonable substitute for the cone penetration test.

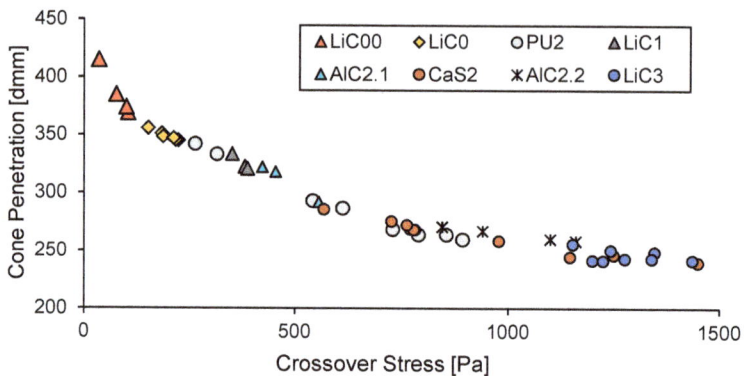

Figure 11. Cone penetration vs crossover stress for various greases.

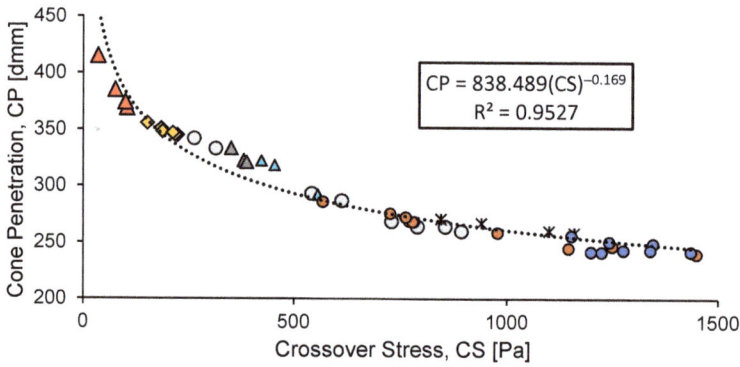

Figure 12. Cone penetration (CP) vs crossover stress (CS) for various grease types with a power fit.

5. Discussion

When considering tests of grease consistency, it is important to realize that consistency is an overall property of a given sample. Even within a sample, there are likely localized differences in consistency that can cause different samples of the same batch to show different results. This can be especially pronounced when a grease has been in storage and/or has experienced large temperature fluctuations. As is displayed in Figure 13, greases can even appear visually nonhomogeneous. The non-homogeneity of the grease by itself can cause a significant difference between an unworked sample and one that is worked for merely 60 strokes in a grease worker.

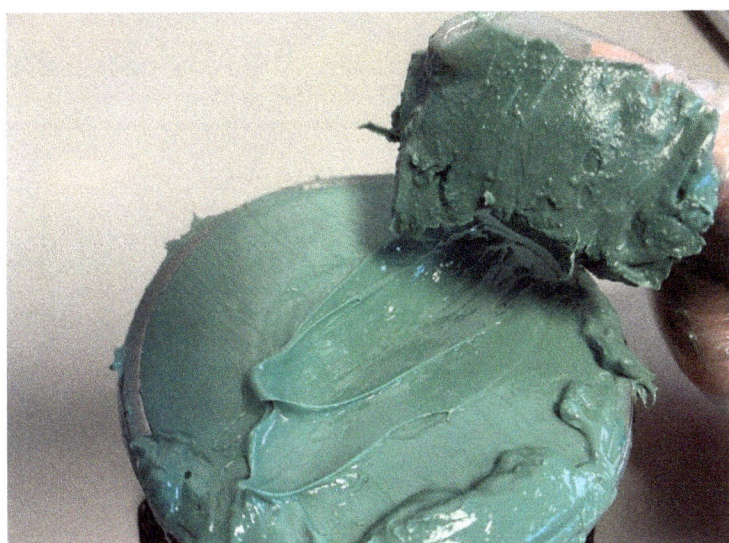

Figure 13. Smooth texture of worked grease in the cup below compared to the rough texture of unworked grease above it.

Though it is important to note that some of the apparent uncertainty of a grease's consistency is due to the grease sample itself, the method for testing consistency plays a dominant role.

5.1. Rheometer Penetration

Overall, the rheometer penetration test shows different trends with different greases. It is possible that the results correlate with some property that was not investigated, but there is no clear correlation between it and any of the other tests considered. Perhaps finding the properties—such as tackiness, base oil viscosity, or other rheological properties—responsible for causing the poor correlation would help give a deeper understanding of the tests themselves as well as grease performance.

Nevertheless, an interesting metaphor between the rheometer penetration test and the cone penetration test can be drawn despite significantly different geometries. The minimum penetration value that corresponds to an NLGI grade (grade 6) is 85 dmm, while the maximum penetration value that corresponds to an NLGI grade (grade 000) is 475 dmm. This is despite the fact that the minimum possible value of penetration is 0 dmm and the maximum value (where the cone hits the bottom of the cup) is 635 dmm. If this is scaled to the rheometer test with a 100 mm/100 starting gap, this corresponds to a minimum penetration of 13.4 mm/100 and maximum penetration of 74.8 mm/100. Despite the completely different geometry, this appears to match somewhat closely with the desirable middle region for the rheometer penetration test. This is especially true for greases within the range of grade 0 to grade 2.

5.2. Cone Penetration

Though the cone penetration remains one of the most common tests, many critics contend that the results of the test do not indicate useful information about a grease. For instance, some point out that it is likely more important to look at pumpability and other grease flow characteristics than cone penetration [26] when considering a grease for a given application. However, the test is such a fundamental tool in measuring a grease that it is unlikely to go away any time soon. In addition, it has an advantage over the other methods used herein in that it can test very firm greases. Greases corresponding to an NLGI grade of 4 or above are generally unsuitable for use in a rheometer, while the cone is designed such that it will give a meaningful penetration value.

A look at the interaction of grease, the cone, and the cup during the cone penetration test shows that there are different phases of the test. The first phase is where the cone penetrates initially and the portion of the cone with a very steep angle causes rapid penetration. This is useful for testing very firm greases where a normal wide cone design would show a negligible penetration.

The second phase involves shallow penetration of the "main" cone body, which is roughly independent of the diameter of the grease cup. According to ASTM D217, this takes place for penetration values below 265 dmm. In this phase, grease begins to be lifted out of the cup, but not enough to significantly impact results.

The third phase is a transition phase, where the grease is lifted out of the cup and begins to be squeezed between the cone and the cup rim. In this region, the geometry of the cup starts to become a significant factor in penetration measurement. This roughly corresponds to the region of penetration between 265 and 400 dmm.

Finally, the fourth phase of cone penetration is where the cup geometry plays a major role in penetration. A large amount of grease is squeezed between the cone and the lip of the cup, causing the cup geometry to play a major role in determining penetration. If a cup of grease with different dimensions were used, it is expected that the penetration values would be significantly different. As is mentioned in ASTM D217, in order to obtain consistent readings for grease within this region (above 400 dmm), it is imperative to center the cone exactly above the cup.

A final note on the cone penetration test is that there are scaled-down alternatives given by ASTM D1403 with correlations between these tests and the full-scale test. These linear correlations are clearly empirical, pointing out some difficulty in describing the behavior of grease during this test with an analytical method. In addition, these tests are further restricted by not allowing the testing of 000 and 00 greases.

5.3. Critical Stress

The evaluation of critical stresses through oscillatory rheometry appears to be a useful tool in measuring grease properties. The yield stress method used appears to be more sensitive to variables such as pre-shear and overfilling compared to the crossover stress method. In addition, the yield stress method does not appear to correlate well with cone penetration. Results from this study as well as the paper defining the procedure [10] show that a grease with a higher cone penetration may or may not have a lower yield stress. This does not mean that this method has no value but does indicate that it is a poor choice for estimating cone penetration and NLGI grade.

6. Conclusions

After a procedure was established for conducting rheometer penetration tests, the method was used to assess the consistency of various greases and compared to other methods of assessing consistency. It was found that the rheometer penetration test does not correlate well with any of the other tests and is, therefore, only recommended to quickly and easily monitor the change in consistency to a particular grease. Though using this method is acceptable to monitor the change in consistency to a particular grease, it is not acceptable to compare the consistency of different greases to each other. There is clearly some parameter responsible for the lack of agreement between the rheometer penetration test and other tests and investigating this lack of agreement can be beneficial for understanding the tests themselves as well as the general nature of grease.

An interesting result from the experiments is that the crossover stress test appears to be an excellent substitute for the cone penetration test. This test is simple to perform, can use a very small sample, is not particularly sensitive to sample preparation, allows for temperature control, and correlates very well with cone penetration. The main disadvantages are that it requires an expensive instrument and relies on arbitrarily chosen parameters such as plate diameter, gap height, and oscillatory frequency. Nevertheless, many researchers use the same variables and if a 25 mm flat plate is used with a 1 mm gap at 1 Hz, the equation given by Figure 11 would be a valid way of estimating cone penetration using the crossover stress. This relationship holds true for all grease types tested and is expected to hold for any other grease type.

Overall, these methods of assessing consistency can be used as a tool for monitoring the degradation of grease due to processes such as shear, oil bleed, contamination, and chemical reactions. However, it is important to keep in mind that the consistency of a grease can appear to change simply because a given sample has a different balance of thickener and oil compared to the average. Thus, appropriate sample selection is an important factor for in-service sampling and even sampling of an unused grease that has been in storage. Once an appropriate sample is taken, it appears that using oscillatory rheometry to calculate the crossover stress has a significant value for reliably assessing the consistency of a grease.

Author Contributions: A.G. is the main author of the paper and received significant guidance, insight, and technical assistance from Professor M.M.K. during writing. All authors have read and agreed to the published version of the manuscript.

Funding: This research received no external funding.

Conflicts of Interest: The authors declare no conflict of interest.

References

1. Lugt, P. A Review on Grease Lubrication in Rolling Bearings. Tribology Transactions—TRIBOL TRANS. 52. 470–480. *Tribol. Trans.* **2009**, *52*, 470–480. [CrossRef]
2. Rezasoltani, A.; Khonsari, M. Mechanical degradation of lubricating grease in an EHL line contact. *Tribol. Int.* **2017**, *109*, 541–551. [CrossRef]
3. ASTM D217-19b. *Standard Test Methods for Cone Penetration of Lubricating Grease*; ASTM International: West Conshohocken, PA, USA, 2019.
4. ASTM D1403-20b. *Standard Test Methods for Cone Penetration of Lubricating Grease Using One-Quarter and One-Half Scale Cone Equipment*; ASTM International: West Conshohocken, PA, USA, 2020.
5. Rezasoltani, A.; Khonsari, M. On the Correlation Between Mechanical Degradation of Lubricating Grease and Entropy. *Tribol. Lett.* **2014**, *56*, 197–204. [CrossRef]
6. Lijesh, K.P.; Khonsari, M. On the Assessment of Mechanical Degradation of Grease Using Entropy Generation Rate. *Tribol. Lett.* **2019**, *67*, 50. [CrossRef]
7. Barnes, H. The yield stress—A review or 'παντα ρει'—everything flows? *J. Non-Newton. Fluid Mech.* **1999**, *81*, 133–178. [CrossRef]
8. Couronné, I.; Vergne, P.; Ponsonnet, L.; Truong-Dinh, N.; Girodin, D. Influence of Grease Composition on its Structure and its Rheological Behavior. In Proceedings of the Leeds-Lyon Conference, Thinning Films and Tribological Interfaces, Leeds-Lyon, UK, 14–17 September 2000.
9. Zakani, B.; Grecov, D. Yield Stress Analysis of a Fumed Silica Lubricating Grease. *Tribol. Trans.* **2018**, *61*, 1131–1140. [CrossRef]
10. Cyriac, F.; Lugt, P.; Bosman, R. On a New Method to Determine the Yield Stress in Lubricating Grease. *Tribol. Trans.* **2015**, *58*, 1021–1030. [CrossRef]
11. Zhou, Y.; Bosman, R.; Lugt, P. A Master Curve for the Shear Degradation of Lubricating Greases with a Fibrous Structure. *Tribol. Trans.* **2019**, *62*, 78–87. [CrossRef]
12. Zhou, Y.; Bosman, R.; Lugt, P. On the Shear Stability of Dry and Water-Contaminated Calcium Sulfonate Complex Lubricating Greases. *Tribol. Trans.* **2019**, *62*, 626–634. [CrossRef]
13. Cyriac, F.; Lugt, P.; Bosman, R. The Impact of Water on the Yield Stress and Startup Torque of Lubricating Greases. *Tribol. Trans.* **2017**, *60*, 824–831. [CrossRef]
14. Lijesh, K.P.; Khonsari, M.; Miller, R. Assessment of Water Contamination on Grease Using the Contact Angle Approach. *Tribol. Lett.* **2020**, *68*, 103. [CrossRef]
15. Rezasoltani, A.; Khonsari, M. On Monitoring Physical and Chemical Degradation and Life Estimation Models for Lubricating Greases. *Lubricants* **2016**, *4*, 34. [CrossRef]
16. Gurt, A.; Khonsari, M. The Use of Entropy in Modeling the Mechanical Degradation of Grease. *Lubricants* **2019**, *7*, 82. [CrossRef]
17. Rezasoltani, A.; Khonsari, M. An engineering model to estimate consistency reduction of lubricating grease subjected to mechanical degradation under shear. *Tribol. Int.* **2016**, *103*, 465–474. [CrossRef]
18. Pierre, N.S. Need to Know: Grease Consistency. Nye Lubricants. Available online: https://www.nyelubricants.com/need-to-know-grease-consistency (accessed on 4 January 2021).
19. Lugt, P. *Grease Lubrication in Rolling Bearings*; John Wiley & Sons: Hoboken, NJ, USA, 2012.

20. Balan, C.; Franco, J. Influence of the Geometry on the Transient and Steady Flow of Lubricating Greases. *Tribol. Trans.* **2001**, *44*, 53–58. [CrossRef]
21. Cardinaels, R.; Reddy, N.; Clasen, C. Quantifying the errors due to overfilling for Newtonian fluids in rotational rheometry. *Rheol. Acta* **2019**, *58*, 525–538. [CrossRef]
22. SKF. *SKF Grease Test Kit TKGT 1: Instructions for Use*; SKF Group: Gothenburg, Sweeden, 2009.
23. Engmann, J.; Servais, C.; Burbidge, A. Squeeze flow theory and applications to rheometry: A review. *J. Non-Newton. Fluid Mech.* **2005**, *132*, 1–27. [CrossRef]
24. Hoffner, B.; Gerhards, C.; Peleg, M. Imperfect lubricated squeezing flow viscometry for foods. *Rheol. Acta* **1997**, *36*, 686–693. [CrossRef]
25. Lee, S.; Denn, M.; Crochet, M.; Metzner, A. Compressive flow between parallel disks: I. Newtonian fluid with a transverse viscosity gradient. *J. Non-Newton. Fluid Mech.* **1982**, *10*, 3–30. [CrossRef]
26. Flemming, W.; Sander, J. Is it Time to Retire the Grease Penetration Test? *NLGI Spokesm.* **2018**, *82*, 14–22.

Article

Evaluating Grease Degradation through Contact Angle Approach

Michael M. Khonsari [1,*], K. P. Lijesh [1], Roger A. Miller [2] and Raj Shah [3]

[1] Department of Mechanical Engineering and Industrial Engineering, Louisiana State University, 3283 Patrick Taylor Hall, Baton Rouge, LA 70803, USA; lijesh_mech@yahoo.co.in
[2] Chemical Engineer, 18342 Char A Banc, Baton Rouge, LA 70817, USA; geauxtigers1@msn.com
[3] Koehler Instrument Company, Holtsville, New York, NY 11742, USA; rshah@koehlerinstrument.com
* Correspondence: khonsari@lsu.edu; Tel.: +1-225-578-9192

Abstract: Grease is highly susceptible to degradation due to regular usage and the severity of the operating conditions. Degradation can negatively impact the performance of grease-lubricated machinery, demanding frequent maintenance to avoid premature failure of machine elements. Quantification of grease degradation has proven to be a formidable task, for which no accepted standards are currently available. In this paper, we extend the results of a novel approach developed recently for the evaluation of the water-resistant property in grease to quantify degradation. The methodology is based on measurements of the contact angle of a water droplet on the surface of a sample of grease. We report the results of extensive tests performed on different grades of lithium complex greases to evaluate the variation of contact angle values with the composition of grease. The measurements were compared with penetrometer readings to quantify a relationship between the grease consistency and contact angle. Detailed study results are also presented on three types of greases sheared in a grease worker for a different number of strokes: contact angle and the yield stress values were measured and compared. Finally, the tribological characteristics were determined for two greases that exhibited a low or high change in their contact angles.

Keywords: grease degradation; contact angle; yield stress; shearing; tribology properties

Citation: Khonsari, M.M.; Lijesh, K.P.; Miller, R.A.; Shah, R. Evaluating Grease Degradation through Contact Angle Approach. *Lubricants* **2021**, *9*, 11. https://doi.org/10.3390/lubricants9010011

Received: 24 December 2020
Accepted: 13 January 2021
Published: 18 January 2021

Publisher's Note: MDPI stays neutral with regard to jurisdictional claims in published maps and institutional affiliations.

Copyright: © 2021 by the authors. Licensee MDPI, Basel, Switzerland. This article is an open access article distributed under the terms and conditions of the Creative Commons Attribution (CC BY) license (https://creativecommons.org/licenses/by/4.0/).

1. Introduction

Grease has widespread application in machinery components such as rolling element bearings, pin bushings and journal bearings [1–7], gears [8–10], slide-ways [11,12], and the like. In these applications, grease composition changes with time, degrading its performance. As a result, the efficiency of the machine deteriorates to the extent that, eventually, the grease can no longer adequately protect the surfaces, at which point failure becomes imminent. To avoid forced shut down, machine operators are required to periodically inspect the health of the grease and replenish or replace it, as deemed necessary.

Grease degradation occurs due to physical changes, chemical changes, or a combination thereof [13–15]. Physical degradation occurs due to bleed-off and/or evaporation of base oil and contamination by particles and/or water. This type of degradation primarily prevails during the shearing of greases below 50 °C. On the other hand, chemical degradation is a result of oxidation of the base oil, or depletion of the additives, occurring at temperatures higher than 50 °C. In general, grease is more prone to physical degradation at high operating speed (i.e., high shearing rate), while chemical degradation occurs at high operating temperatures or during long-term storage [15]. The focus of the present work is on assessing the degradation of grease due to physical change.

The industry typically measures the physical changes in grease by evaluating its consistency through the worked penetration test provided in the American Society for Testing and Materials (ASTM) standard D217 [16]. In this test, a cone of standard shape and weight is released to fall into a cup of grease, after which the depth of penetration into the sample is recorded. The larger the penetration value, the lower is the consistency, and vice versa. ASTM D217, however, requires a large amount of grease, which is not practical when trying to assess the consistency of small amounts of grease taken from roller element bearings or other lube points. To overcome this complication, Rezasoltani and Khonsari [17] employed a rheometer for assessing grease degradation by monitoring the change in the rheological properties and correlating it to the mechanical degradation of the grease. The mechanical degradation was determined by squeezing the grease between two parallel plates and measuring the difference in the plate position after 60 s. The mechanical degradation of the grease can also be determined in a rheometer by measuring the yield stress, zero viscosity, cross-over stress, etc.

Lijesh and Khonsari [18] extended the approach proposed by Rezasoltani and Khonsari [17] for developing a predictive model for determining the degradation of grease from their operating NLGI grade and thereafter estimating their remaining useful life. Specifically, they based their degradation assessment on a relationship between the change in the grease consistency and entropy generation. For example, the drop in the consistency of a pristine grease of NLGI grade 2 to NLGI grade 1 or 0 can be considered an indication of the reduction of performance and the necessity for re-lubrication. Testing via a rheometer requires far less grease compared to D217; however, an expensive rheometer and appropriate technical expertise are required, making it unaffordable for many industries.

Lijesh et al. [19] very recently developed a unique approach to quantify the water repellant properties of grease by measuring the contact angle of a droplet of water on the surface of a grease sample. In this method, a small quantity of grease is spread over a surface, a water droplet is dropped on it, and the contact angle of the water droplet is measured. The contact angle values are dependent on the type and composition of the grease, i.e., thickener, base oil, and additives. Thus, we hypothesize that the degradation of grease can also be effectively characterized using the contact angle approach. The immediate advantage of this approach is that only a small quantity of grease is needed. Ideally, a portable instrument can be built for testing grease performance in the field [20,21].

To validate the hypothesis, experiments were performed by degrading grease in a grease worker and measuring the contact angle after periodic intervals. In the present work, the evaluation was performed for three types of greases. The change in the contact angle values after a different number of strokes was considered for evaluating the degradation characteristics of the grease. To corroborate the findings, the same greases were also tested in a rheometer. To gain further confidence, two of the greases rendering higher and lower variation in contact angle with time were tested for tribological performance in a tribometer.

The outline of this paper is as follows. Section 2 provides the details of the instruments used for shearing grease, i.e., a grease worker for measuring the rheological properties and a rheometer for measuring contact angles. Section 3 is devoted to the presentation of results, followed by a discussion in Section 4. In Section 5, summary and concluding remarks are provided.

2. Materials and Methods

Table 1 shows the list of eight commercially available greases considered for the present investigation. It includes the base oil, the type of thickener, the color, and NLGI grades of each grease.

Table 1. Grease designation with thickener and base oil types.

Grease Label	Thickener Types	Base Oil Type	Color	NLGI Grade
Li_m	Lithium based	Mineral oil	Blue	3, 2, 1, 0, 00
Li_P	Lithium based	Poly-alpha-olefin oil	Pink	2
CaS	Calcium sulfonate	Mineral oil	Green	2
PU	Poly-urea	Mineral oil	Blue	2

2.1. Grease Worker

The grease was degraded by shearing in the grease worker rig shown in Figure 1. This rig contained a 2 hp gear reduction motor, plunger assembly, a grease cup, cover, and an electrical counter. The plunger assembly consisted of a handle, a shaft, and a perforated plate. The handle had an oval shape slot to convert the rotary motion of the motor to linear-reciprocating motion. The number of strokes was counted using an electric counter. The shearing action was induced in grease by reciprocating the handle and shaft inside the grease cup and forcing grease to pass through a series of holes in the plunger.

Figure 1. Modified grease worker.

Grease samples of 30 g by weight were sheared in the grease cup at $1\ \text{s}^{-1}$ shear rate at room temperature (25 °C). The grease from the cup was then used for evaluation under three cases: Case 1: 10,000 strokes, Case 2: after 86,400 strokes (i.e., after 24 h), and Case 3: after 172,800 strokes (i.e., after 48 h). Measurements of grease samples were performed using a digital scale with an accuracy of 0.1 mg. The testing conditions were selected such that the grease sheared enough to show considerable degradation within the cases.

2.2. Water Droplet Analyzer

The drop shape analyzer (Krüss, Hamburg, Germany) shown in Figure 2 was used to determine the contact angle of the water droplets on the grease surface. This setup consisted of a camera (IDS UI-5480CP-M-GL GigE camera) and adjustable lens (Thorlabs AC254-075-A-ML Lens) through which the water droplet on grease sample was analyzed. Using an adjustable screw, the height of the sample stage was adjusted such that the water droplet was in line with the lens height. The angle of the lens was further adjusted by an alignment screw, as needed. The apparatus provided a monochromatic blue light that helped in obtaining a clear and distinguishable image of the droplet from the background. The apparatus used the captured image to calculate the contact angle θ via the built-in software.

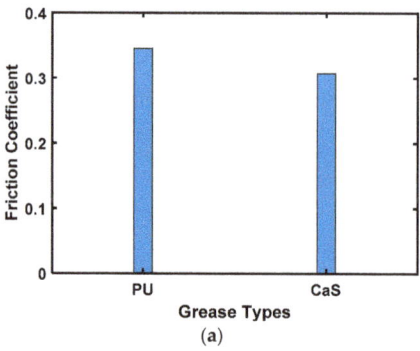
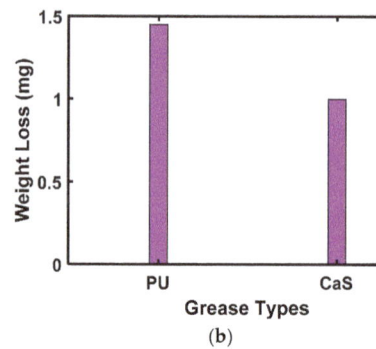

Figure 2. Drop shape analyzer.

A mold made of polymer with rhombus shapes was used in the present work to achieve consistent thickness and a uniform surface of grease. To remove the trapped air, the grease in the mold slots was completely compressed. The grease sample was refilled again if trapped air was observed. The slot containing grease to be tested was kept in front of the camera and a 5 µL water drop was placed on the grease surface using a 10 µL syringe. Due to the semi-solid nature and complex structure of grease, the dimensions of the droplets on the grease surface changed with time, making it difficult to capture the image instantaneously. To address this issue, a video of the droplet was recorded for more than 5 min at 3 frames/s. Images of the water droplet from the video after 60 s were considered for measuring the contact angles. The standard operating procedure is explained in Appendix A and also an explanation of the methodology is provided elsewhere [19]. It was made sure that the grease sample considered for testing was at 25 °C.

2.3. Rheometeric Tests

The yield stress results for grease sheared for different numbers of strokes were measured using a rheometer (Anton Paar MCR 301, Graz, Austria) shown in Figure 3. The details of the specification of the rheometer are provided elsewhere [18].

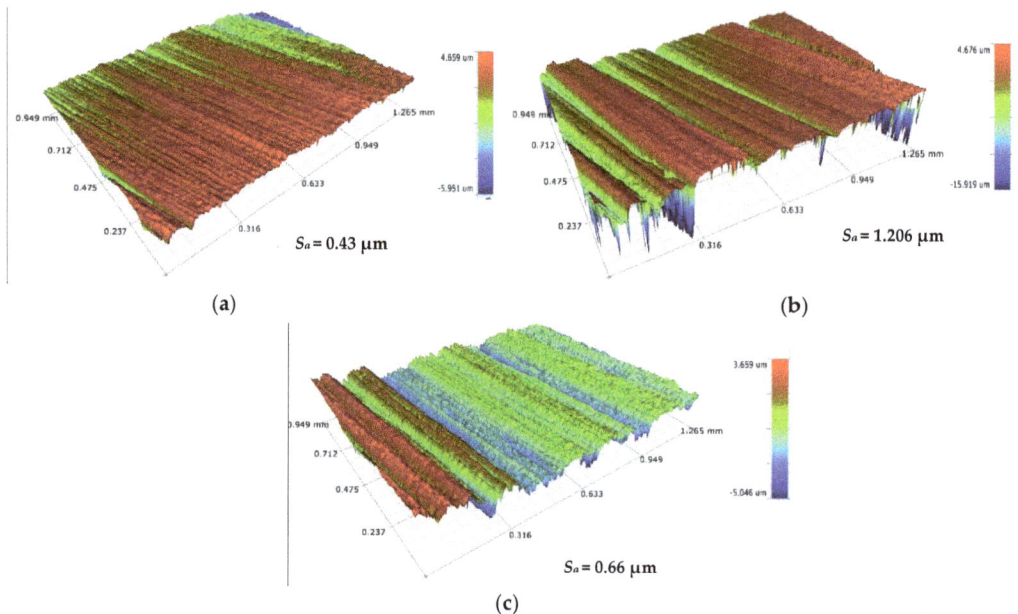

Figure 3. Rheometer for measuring yield stress and penetration.

To determine the yield stress, a grease sample of 2 mm thick and 15 mm diameter was placed onto the stationary surface (see Figure 4a). The plate was moved to the desired gap thickness of 1 mm (Figure 4b) and the excess grease was trimmed off (Figure 4c). To remove the deformation history due to the squeezing of grease, sufficient rest time was provided to relax.

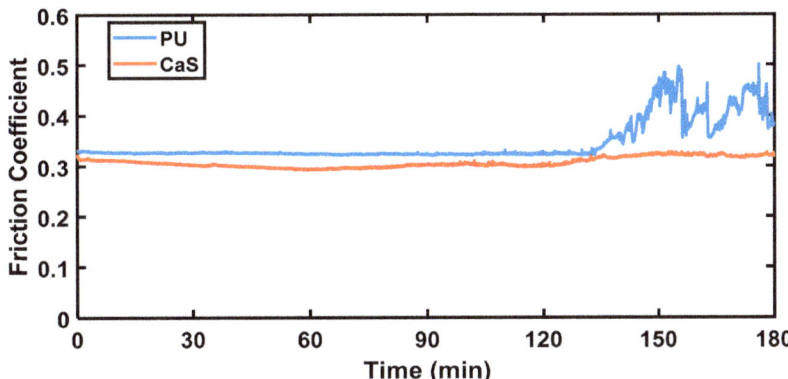

Figure 4. Procedure for placing the grease samples between the plate and stationary surface: (**a**) place the grease sample; (**b**) squeezing the grease to the required gap; (**c**) trimming of the excess grease.

The yield stress values were determined from the shear stress-strain plot. The plot was obtained by oscillate-sweeping the plate from 0.001% to 1% at a fixed frequency of 1 Hz. The oscillatory strain sweep approach was adopted due to its robustness and reliability. Further, the results obtained were insensitive to the geometry of the plates, surface roughness, the gap between the plates, and the frequency of shearing [22,23]. The yield stress is the point on the stress-strain curve where the coefficient of determination (R^2) between the third-order polynomial fit of the experimental values and a linear fit is found to be greater than 99.5% [23].

2.4. Tribometer

The change in the tribological performance (friction and wear) of the grease with water contamination was studied using a tribometer setup, shown in Figure 5. The setup consisted of a stationary disk with projection sliding against a rotating disk with a groove (see Figure 5). The tribometer setup was designed for performing pin-on-disk; However, to perform disk-on-disk tests, a disk holder was designed and built (see Figure 5).

The stationary disk was fastened to the disk holder and connected to a dual load sensor through the suspension. Similarly, the rotating disk was fastened to a lubricating cup and then to the driven pulley. The driven pulley was connected to the motor using the driving pulley through a belt-pulley drive system. The dual load sensor was attached to the motion control drive to measure the applied load in the vertical direction and frictional force in the horizontal direction. Using the motion control drive, the required displacement in the vertical as well as horizontal motion was achieved. Inbuilt software was used to control the motions of the motor and control drive. The additional details of the setup are available in [24,25].

The mean diameter and thickness in the stationary part were 30 and 2 mm, respectively. The depth, mean diameter, and thickness of the groove in the rotating part were 1, 30, and 8 mm, respectively. Experiments were performed at 30 N load and 0.126 m/s for 3 h. The weight loss was determined using a weighing gauge having an accuracy of 0.1 mg.

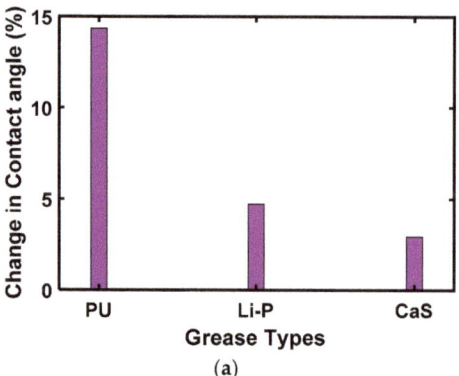
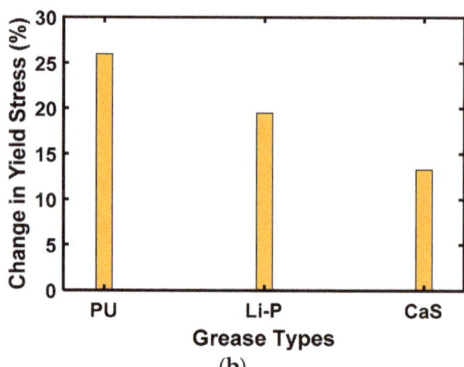

Figure 5. Tribometer setup.

3. Results

3.1. Pristine Grease

Initial testing was performed on the pristine grease to study the variation in contact angle with different grades of lithium-based greases. Greases were sheared in the grease worker for 60 strokes, and the worked penetration values were determined using a penetrometer. Figure 6 shows the results of the contact angle measurement as a function of the worked penetration. From this figure, the existence of a linear correlation between different NLGI grade greases and contact angle values can be observed.

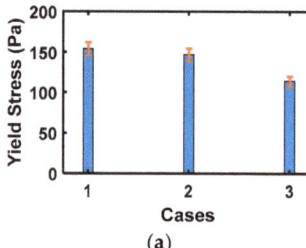
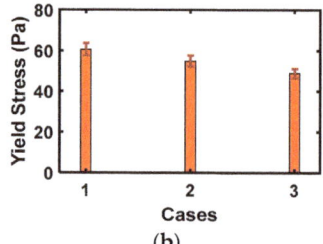
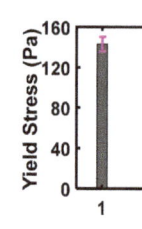

Figure 6. Worked penetration vs. contact angle for different NLGI grade grease.

3.2. Degraded Grease

In this section, results are provided for degraded greases induced in the grease worker for different strokes. The contact angle values are measured and compared with the rheological property.

3.2.1. Water Droplet Analyzer

A water droplet was dispensed on the grease surface. The droplet was recorded for more than 60 s. From the recorded video, the images of the water droplet at 60 s were considered for evaluation. The contact angle of the captured images was determined using the built-in software. The values of the contact angle at 60 s for PU grease for different cases are plotted in Figure 7. It can be observed that the contact angle values tended to drop with the increase in the number of strokes.

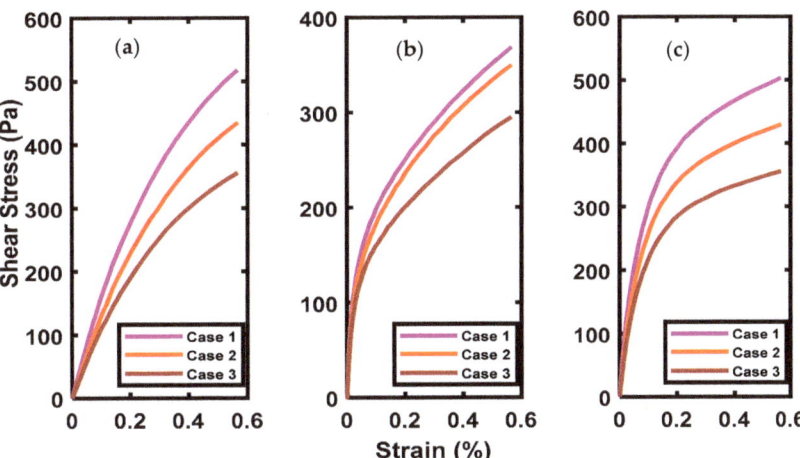

Figure 7. Contact angle for PU grease for different cases.

Similarly, the contact angle values obtained for Li-P and CaS greases for different cases are shown in Figure 8a,b and Table 2. Referring to these figures, it can be inferred that for both types of greases, the contact angle values were reduced with the number of strokes.

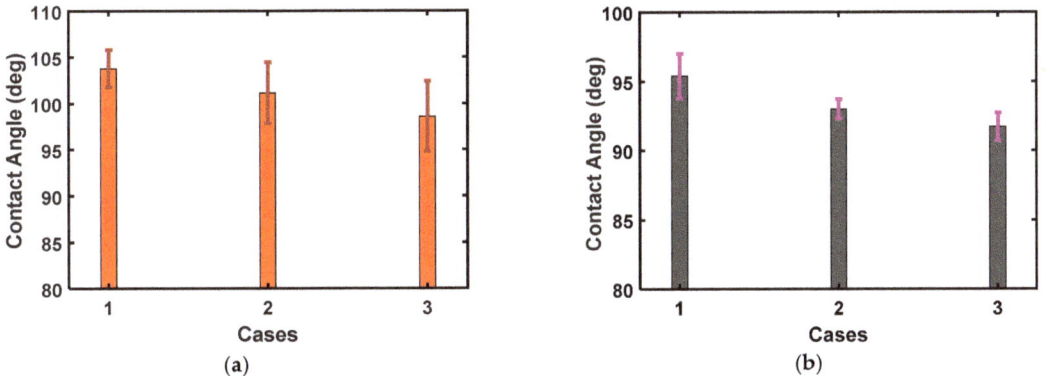

Figure 8. Contact angle for Li-P and CaS greases for different cases: (**a**) Li-P; (**b**) CaS.

Table 2. Average contact angle values for Li-P and CaS greases.

Cases	Li-P	CaS
1	103	95.4
2	101	93
3	98.6	91.7

3.2.2. Rheometric Measurements

The yield stress values for all three types of greases and for Cases 1–3 were determined from the shear stress-strain plots obtained from the oscillatory sweep experiments [26,27]. The shear stress-strain plots for PU, Li-P, and CaS greases are presented in Figure 9a–c, respectively.

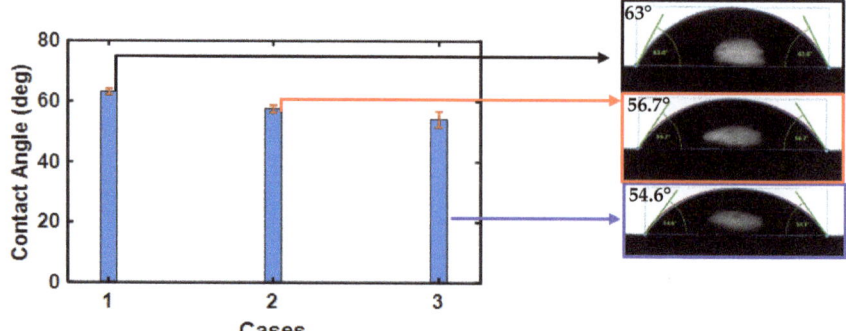

Figure 9. Shear stress versus shear rate plot for Cases 1–3, for all greases: (**a**) PU; (**b**) Li-P; (**c**) CaS.

From the obtained shear stress-strain values and using the linear fit function in Matlab, the yield stress values were determined for $R^2 > 99.5\%$. Yield stress values for Cases 1–3 were determined for PU, Li-P, and CaS greases and the values are shown in Figure 10a–c, respectively. It is observed that the yield stress values were reduced with the increasing number of strokes.

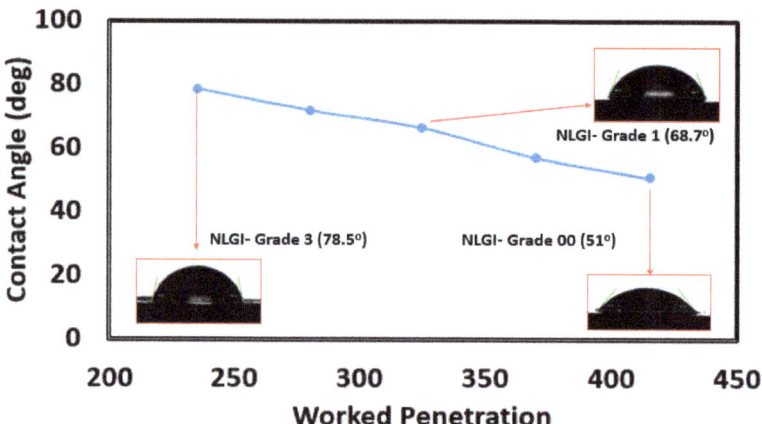

Figure 10. Yield stress for PU, Li-P, and CaS greases for different cases; (**a**) PU; (**b**) Li-P; (**c**) CaS.

4. Discussion

The contact angle values of water droplets on grease surfaces depend on the composition and types of the ingredients used for developing greases. Therefore, it was postulated that greases having the same types of ingredients, but in different proportions, have different contact angles. To test the hypothesis, contact angle values were measured for different grades of Li-m greases, and the results are provided in Figure 6. A linear relationship between the contact angle values and different NLGI grade penetration values was observed. This concludes that through the contact angle values, different grades of the same kind of grease can be determined. Further, in a very recent finding, Lijesh and Khonsari [18] demonstrated the existence of a relationship between the NLGI grade reduction and degradation. Therefore, by measuring the contact angle of the grease, one can determine the degradation of grease.

The three types of greases (PU, Li-P, and CaS) were degraded in the grease worker, and the contact angle values were measured for Cases 1–3. The measured contact angle values for PU grease are shown in Figure 7 and for Li-P and CaS greases are shown in

Figure 8a,b, respectively. From these figures, it can be observed that the contact angle values decreased with an increasing number of strokes, i.e., with the degradation of grease, the contact angle values reduced. The percentage changes of contact angle values between Cases 1 and 3 for all three greases are plotted in Figure 11a.

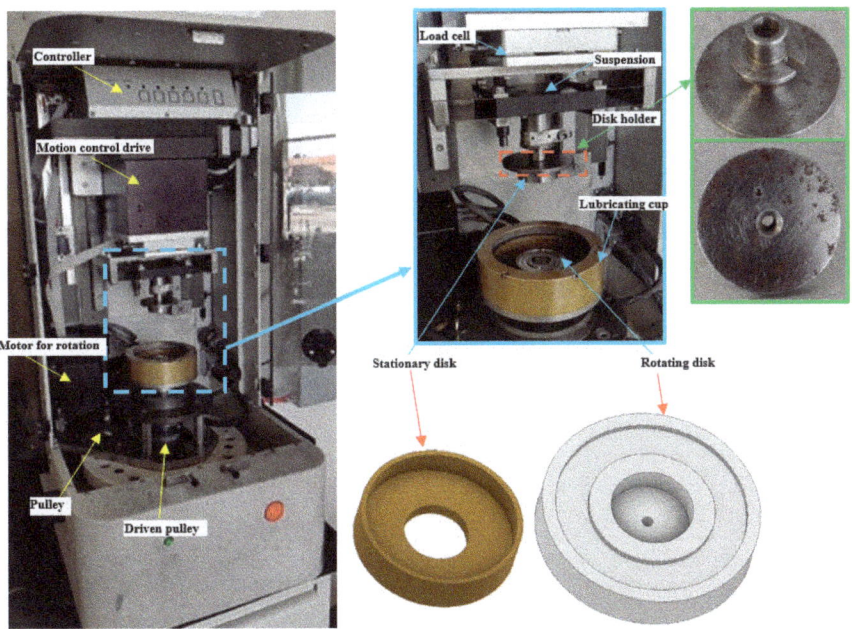

Figure 11. Change in contact angle and yield stress values calculated between Cases 1 and 3: (**a**) contact angle; (**b**) yield stress.

Further, to gain more confidence in the results obtained using the contact angle approach, the rheological change in grease with degradation was evaluated by determining yield stress values. Shear stress versus strain values for all greases and cases are plotted in Figure 9, and the measured yield stress values for PU, Li-P, and CaS greases are shown in Figure 10a–c. The change in yield stress values was determined for all the greases and plotted in Figure 11b. From Figure 11a,b, the highest and lowest changes were determined for PU and CaS greases, respectively. In other words, the PU grease degraded faster while CaS degraded slowly.

Tribological Performance

To further assess the importance of a grease having lower degradation with time, experiments were performed on a tribometer for 3 h, and tribological properties (i.e., friction coefficient and weight loss) were determined for Pu and CaS greases. The online friction coefficient values for both greases are plotted in Figure 12. In the case of CaS grease, the friction coefficient values remained almost constant, signifying satisfactory operation. For the PU grease, however, the friction coefficient values exhibited significant fluctuation after 130 min. The reason for the fluctuation can be attributed to the interference of worn-out particles between the sliding surfaces. To investigate, an optical surface profilometer was used to examine the 3D profiles of the rotating disk surface. Figure 13a shows the surface profile at the pristine condition and Figure 13b,c shows the profiles for the PU and CaS greases, respectively. The measured roughness value of the pristine surface was 0.43 μm, while the surfaces tested with PU and CaS greases were determined to be 1.2 and 0.66 μm, respectively. From the surface roughness values and the 3D profile images provided in

Figure 13, the following observations are made: (i) the surface roughness values increased by performing experiments, and (ii) the surface of rotating disks operated with PU grease caused more damage than CaS grease. Therefore, the CaS grease protected the disk surfaces from wearing out, whereas the PU grease degraded faster and after 130 min. of shearing, it failed to protect the surfaces.

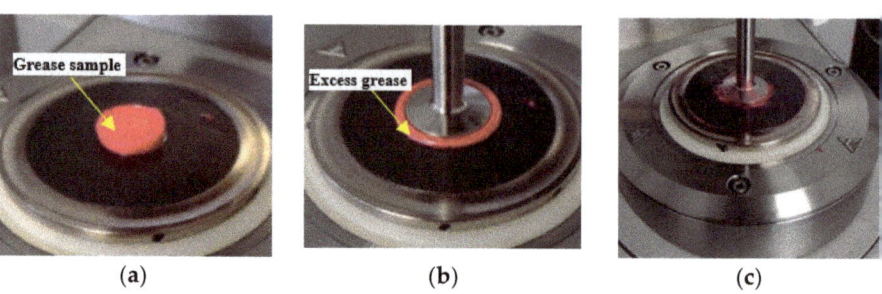

Figure 12. Online recorded friction coefficient values for PU and CaS greases.

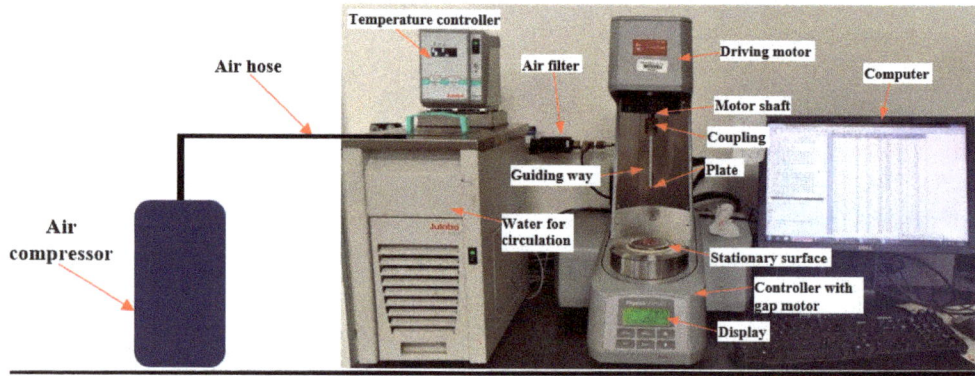

Figure 13. 3D profile image of rotating disk: (**a**) pristine surface; (**b**) worn-out surface with PU grease; and (**c**) CaS grease.

The average friction coefficient values and measured weight loss after completion of the experiment are plotted in Figure 14a,b, respectively. From these figures, it can be concluded that CaS showed better tribological performance than PU grease. This finding is in agreement with the findings of the contact angle approach.

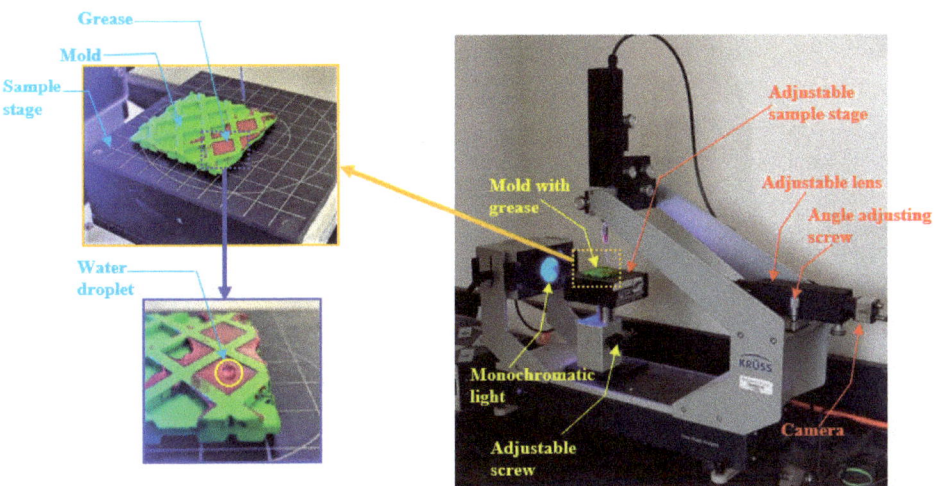

Figure 14. Average friction coefficient and weight loss values for PU and CaS greases: (**a**) average friction coefficient; (**b**) weight.

5. Conclusions

This paper is dedicated to the development of a procedure for the characterization of grease degradation using the contact angle approach. In this approach, a water droplet is placed onto the grease surface and the contact angle of the water droplet is measured. It is postulated that this approach has the potential to reflect the change in the composition of the grease, which in turn can be employed for determining the degradation of the grease. To validate, a contact angle test was performed on different grades of Li-m greases and the results were compared with those obtained by penetration tests. A linear correlation was observed between them, indicating the capability of the contact angle approach in determining the change in consistency of the grease. PU, Li-P, and CaS greases were sheared in a grease worker for a different number of strokes and their degradation was evaluated by measuring the contact angle values and comparing these with the yield stress values. From both approaches, it was found that the PU grease degraded faster and CaS grease degraded slower. To assess the importance of the finding, experiments were performed on a tribometer to evaluate the tribological properties of PU and CaS greases. From the experimental results, it was concluded that the wear and friction coefficient values for PU grease were higher than those of CaS grease. Further, for PU grease, the generation of wear particles was observed. This concludes that CaS grease degraded at a lower rate compared to PU grease, which is in agreement with the findings of the contact angle approach.

Author Contributions: Conceptualization, M.M.K. and K.P.L.; Methodology, software, validation, formal analysis, investigation: K.P.L.; resources, data curation, writing—original draft preparation, M.M.K. and K.P.L.; writing—review and editing, M.M.K., R.A.M. and R.S.; visualization, supervision, project administration, funding acquisition: M.M.K. All authors have read and agreed to the published version of the manuscript.

Funding: M. M. Khonsari and K. P. Lijesh gratefully acknowledge the support of background research on this subject through the LIFT^2 Program, grant number LSU-2021-LIFT-007.

Institutional Review Board Statement: Not applicable.

Informed Consent Statement: Not applicable.

Data Availability Statement: Data will be provided when requested.

Conflicts of Interest: The authors declare no conflict of interest.

Appendix A

Step 1: Collect grease samples from the machinery or components which need to be tested. The volume of sample required is 75 mm^3; i.e., around 0.25 g of grade 2 grease.

Step 2: Apply the grease sample to the rhombus slot.

Step 3: Remove the excess grease. The grease is reapplied if air bubbles are entrapped in the grease.

Step 4: Keep the mold in front of the lens of the drop shape analyzer.
Step 5: Start button in the software provided for drop shape analyzer is clicked.
Step 6: Drop water droplet on the grease surface.

Step 7: Wait for more than 60 s.
Step 8: Stop the recording.
Step 9: Time for water droplet touching the grease surface is noted and the contact angle of the droplet after 60 s from the noted time is measured.

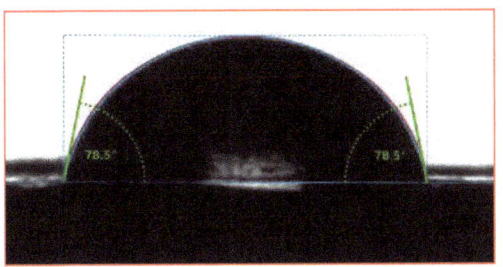

References

1. Cobb, L.D. Grease Lubrication of Ball Bearings. *Lubr. Eng.* **1952**, *8*, 120–124.
2. Kawamura, T.; Minami, M.; Hirata, M. Grease life prediction for sealed ball bearings. *Tribol. Trans.* **2001**, *44*, 256–262. [CrossRef]
3. Baker, A.E. Grease Bleeding—A Factor in Ball Bearing Performance. *Nlgi Spokesm.* **1958**, *22*, 271–279.
4. Lundberg, J.; Berg, S. Grease Lubrication of Roller Bearings in Railway Wagons. Part 2: Laboratory Tests and Selection of Proper Test Methods. *Ind. Lubr. Tribol.* **2000**, *52*, 76–85. [CrossRef]
5. Cann, P.M.; Doner, J.P.; Webster, M.N.; Wikstrom, V. Grease Degradation in Rolling Element Bearings. *Tribol. Trans.* **2001**, *44*, 399–404. [CrossRef]
6. Booser, E.R.; Khonsari, M.M. *Grease and Grease Life, Encyclopedia of Tribology*; Chung, Y.W., Wang, Q.J., Eds.; Springer: New York, NY, USA, 2013; pp. 1555–1561.
7. Khonsari, M.M.; Booser, E.R.; Miller, R. How Grease Keep Bearings Rolling. *Mach. Des.* **2010**, *82*, 54–59.
8. Fukunaga, K. Allowable surface durability in grease lubricated gears. *Tribol. Trans.* **1988**, *31*, 454–460. [CrossRef]
9. Krantz, T.L.; Robert, F.H. A study of spur gears lubricated with grease-observations from seven experiments. In Proceedings of the 58th Meeting of the Society for Machinery Failure Prevention Technology (MFPT) NASA, Virginia Beach, VA, USA, 26–30 April 2004; pp. 1–15.
10. Schultheiss, H.; Stemplinger, J.P.; Thomas, T.; Karsten, S. Influences on Failure Modes and Load-Carrying Capacity of Grease-Lubricated Gears. In Proceedings of the International Conference on Gears, Munich, Germany, 5–7 October 2015; pp. 5–7.
11. Ming, Z.Y.H.Y.Z. Development of Slide-way Sealing Grease 7035 for Sintering Machine. *Synth. Lubr.* **2006**, *1*, 5.
12. Aranzabe, A.; Estibaliz, A.; Arrate, A.; Raquel, F.; Jesus, T.; Jo, A.; Raj, S. Comparing different analytical techniques to monitor lubricating grease degradation. In *NLGI Spokesman-Including NLGI Annual Meeting-National Lubricating Grease Institute*; NIGL: Kansas City, MO, USA, 2006; Volume 70, pp. 17–30.
13. Ito, H.; Masami, T.; Toshiro, S. Physical and chemical aspects of grease deterioration in sealed ball bearings. *Lubr. Eng.* **1988**, *44*, 872–879.
14. Rezasoltani, A.; Khonsari, M. On monitoring physical and chemical degradation and life estimation models for lubricating greases. *Lubricants* **2016**, *4*, 34. [CrossRef]
15. Gurt, A.; Khonsari, M.M. The use of entropy in modeling the mechanical degradation of grease. *Lubricants* **2019**, *7*, 82. [CrossRef]
16. ASTM D217-17. *Standard Test Methods for Cone Penetration of Lubricating Grease*; ASTM International: West Conshohocken, PA, USA, 2017.
17. Rezasoltani, A.; Khonsari, M.M. On the Correlation between Mechanical Degradation of Lubricating Grease and Entropy. *Tribol. Lett.* **2014**, *56*, 197–204. [CrossRef]
18. Lijesh, K.P.; Khonsari, M.M. On the assessment of mechanical degradation of grease using entropy generation rate. *Tribol. Lett.* **2019**, *67*, 50. [CrossRef]
19. Lijesh, K.P.; Khonsari, M.M.; Miller, R.A. Assessment of Water Contamination on Grease Using the Contact Angle Approach. *Tribol. Lett.* **2020**, *68*, 1–12. [CrossRef]
20. Williams, D.L. Portable Contact Angle Measuring Device. U.S. Patent 9,958,264, 1 May 2018.
21. Williams, D.L. Portable Contact Angle Measuring Kit. U.S. Patent 9,874,528, 23 January 2018.
22. Cyriac, F.; Lugt, P.M.; Bosman, R. On a new method to determine the yield stress in lubricating grease. *Tribol. Trans.* **2015**, *58*, 1021–1030. [CrossRef]
23. Zhou, Y.; Bosman, R.; Lugt, P.M. A model for shear degradation of lithium soap grease at ambient temperature. *Tribol. Trans.* **2018**, *61*, 61–70. [CrossRef]

24. Lijesh, K.P.; Khonsari, M.M. Characterization of multiple wear mechanisms through entropy. *Tribol. Int.* **2020**, *152*, 106548. [CrossRef]
25. Lijesh, K.P.; Khonsari, M.M. Characterization of abrasive wear using degradation coefficient. *Wear* **2020**, *450*, 203220. [CrossRef]
26. Paszkowski, M.; Olsztyńska, J.S. Grease thixotropy: Evaluation of grease microstructure change due to shear and relaxation. *Ind. Lubr. Tribol.* **2014**, *66*, 223–237. [CrossRef]
27. Paszkowski, M.; Olsztyńska-Janus, S.; Wilk, I. Studies of the kinetics of lithium grease microstructure regeneration by means of dynamic oscillatory rheological tests and FTIR–ATR spectroscopy. *Tribol. Lett.* **2014**, *56*, 107–117.

Technical Note

Formulation to Calculate Isothermal, Non-Newtonian Elastohydrodynamic Lubrication Problems Using a Pressure Gradient Coordinate System and Its Verification by an Experimental Grease

Kunihiko Kakoi

TriboLogics Corporation, Tokyo 110-0016, Japan; kakoi@tribology.co.jp

Abstract: This paper presents a formulation of point contact elastohydrodynamic lubrication analysis for an isothermal, non-Newtonian flow. A coordinate system of the pressure gradient was employed herein. A Couette flow and a Poiseuille flow were considered along the directions of the zero and non-zero pressure gradients, respectively. The Poiseuille flow velocity was assumed to be represented by a 4th-order polynomial of z along the film thickness direction. The Couette flow velocity was assumed to be represented by a linear function of z. Subsequently, the modified Reynolds equation, which contains an equivalent viscosity, was obtained. Using Bauer's rheological model, the formulation presented in this study was applied to a grease that has been previously experimented upon. The results of previous studies were compared with those of the present study and a reasonable agreement was noted. The distribution of the equivalent viscosity showed a notable difference from that of Newtonian flow. The formulation can be incorporated easily to the usual elastohydrodynamic lubrication calculation procedure for Newtonian flow. The method can be easily applied to other non-Newtonian rheological models. The equivalent viscosity can be calculated using the one-parameter Newton-Raphson's method; as a result, the calculation can be performed within a reasonable time.

Keywords: elastohydrodynamic lubrication; isothermal; non-Newtonian; point contact; grease lubrication; Bauer's model; pressure gradient; equivalent viscosity

Citation: Kakoi, K. Formulation to Calculate Isothermal, Non-Newtonian Elastohydrodynamic Lubrication Problems Using a Pressure Gradient Coordinate System and Its Verification by an Experimental Grease. *Lubricants* **2021**, *9*, 56. https://doi.org/10.3390/lubricants9050056

Received: 16 February 2021
Accepted: 11 May 2021
Published: 14 May 2021

Publisher's Note: MDPI stays neutral with regard to jurisdictional claims in published maps and institutional affiliations.

Copyright: © 2021 by the author. Licensee MDPI, Basel, Switzerland. This article is an open access article distributed under the terms and conditions of the Creative Commons Attribution (CC BY) license (https://creativecommons.org/licenses/by/4.0/).

1. Introduction

Performing experiments on non-Newtonian flows, including grease flows, is considerably time-consuming and costly. Therefore, it is important to numerically analyze the phenomena corresponding to non-Newtonian flows. Numerical approaches can help obtain a variety of data that cannot be determined experimentally. Grease flows can be well defined using Bauer's model; however, owing to the extreme complexity of this model, it is difficult to determine the exact solution as well as approximate solutions for point contact, isothermal, non-Newtonian elastohydrodynamic lubrication (EHL) analyses. Therefore, it is convenient that the non-Newtonian EHL calculation can be executed within a reasonable calculation time and without large modification to the usual Newtonian EHL calculation procedure. Kochi et al. [1] performed experiments on grease under soft EHL conditions and measured the film thickness and traction forces. The method proposed in the present study can be applied to the grease considered in Kochi et al. [1] to validate this theoretical approach.

1.1. Classification of Calculation Methods

As shown in Figure 1, the Z-direction is considered to be the film thickness direction. The flow velocities along the X- and Y-directions are denoted by u and v, respectively, which are functions of x, y, and z; however, when considering only z dependency, the velocities can be expressed as $u(z)$ and $v(z)$, respectively. The numerical methods for isothermal, non-Newtonian EHL analyses can be classified in terms of the accuracy of $u(z)$ and $v(z)$, as follows:

Method 1: Exact solution of $u(z)$ and $v(z)$ is obtained.
Method 2: Approximate solution of $u(z)$ and $v(z)$ is obtained.

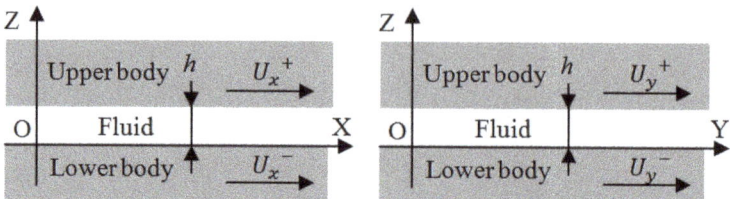

Figure 1. Global coordinate system O, X, Y, Z.

Method 2 can be further classified in terms of the employed coordinate system.
Method 2-A: Local X-direction is the sliding direction.
Method 2-B: Local X-direction is the direction of the pressure gradient.

1.2. Previous Works

Several methods have been proposed to solve isothermal, non-Newtonian EHL problems. Kauzlarich and Greenwood [2] obtained the exact solution of line contact EHL problems considering the Herschel–Bulkley model. Conry et al. [3] obtained an exact solution of line contact EHL problems by considering Eyring's model. Specifically, the velocity $u(z)$ was represented by a function containing cosh, indicating that, in general, $u(z)$ and $v(z)$ cannot be precisely represented using polynomials of z. Dong and Qian [4] obtained an approximate solution of line contact EHL problems considering Bauer's model and using the weighted residual method. Peiran and Shizhu [5] proposed a method to obtain an exact solution of point contact EHL problems for general rheological models by using an equally divided Z-direction mesh. Subsequently, the authors applied the method to a line contact EHL problem. Kim et al. [6] attempted to obtain an exact solution of point contact EHL problems by considering Eyring's model. Ehret et al. [7] considered the X-direction to align with the sliding direction and obtained an approximate solution of $u(z)$ and $v(z)$ represented by the 2nd order polynomial of z by using the perturbation method. Thus, researchers have obtained two effective viscosities: one along the sliding direction and another along the perpendicular direction.

Greenwood [8] focused on the considerable amount of computation time required to obtain an exact solution and compared two approximation methods. Sharif et al. [9] considered the X-direction to align with the sliding direction and developed a method to obtain an exact solution of point contact EHL problems for an arbitrary rheological model. Using this approach, the authors obtained two effective viscosities along the X- and Y-directions. Liu et al. [10] formulated a method to obtain an exact solution of point contact thermal EHL problems considering Eyring's model. Yang et al. [11] formulated a general Reynolds equation for point contact EHL problems by dividing the flow into Couette and Poiseuille flows. Subsequently, the authors obtained an exact solution of line contact EHL problems considering the power law model and demonstrated the effectiveness of their proposed method. Furthermore, Bordenet et al. [12] obtained an exact solution of pure rolling point contact EHL problems considering Bauer's model for $n = 1/2$ and applied the approach to grease.

2. Overview of the Proposed Method

In this work, an isothermal, non-Newtonian EHL formulation considering Method 2-B was developed. Although this approach does not yield the exact solution of $u(z)$ and $v(z)$, the calculation is simple and fast. As shown in Figure 2, the local Xc-direction is considered as the direction of the pressure gradient *d*, and the Yc-direction is considered to be perpendicular to Xc. The flow velocities in the Xc- and Yc-directions are denoted as u_c and v_c, respectively. As the pressure gradient toward Yc is zero, the flow v_c is assumed to be a Couette flow, which can be represented using a linear function of z. As the pressure gradient toward Xc is generally non-zero, the flow u_c is assumed to be a Poiseuille flow, which can be represented using a 4th-order polynomial of z. As u_c and v_c cannot be precisely represented by polynomials of z, they are expanded using polynomials of z. In such cases, the 6th-order or even higher order polynomials can be considered; however, in this work, a lower 4th-order polynomial was employed. A viscosity corresponding to the Newtonian flow was obtained and termed as the equivalent viscosity. To replace the viscosity of the Newtonian flow with the equivalent viscosity, which is a typical process when evaluating EHL problems, the method proposed by Venner and Lubrecht [13] can be used without any change for the isothermal, non-Newtonian EHL calculation. Given a rheological model, any non-Newtonian isothermal point contact EHL problem can be solved using the proposed method.

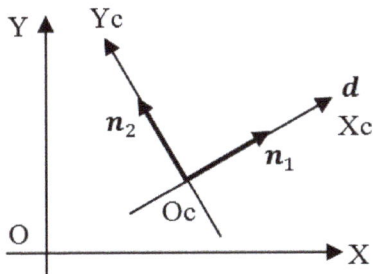

Figure 2. Local coordinate system Oc, Xc, Yc.

The calculation procedure consists of two steps. First, the Newtonian EHL calculation is executed for the base oil that yields the Newtonian pressure distribution. Second, using the pressure distribution as the initial value, the non-Newtonian EHL calculation is executed for the non-Newtonian flow. The local coordinate system, Oc, Xc, and Yc, at a particular point, depends on the pressure gradient and it is determined simultaneously in the process to obtain the pressure distribution. The equivalent viscosity is calculated based on the local coordinate system in each iteration loop to obtain the pressure distribution. Here, the method was applied to an experimental grease characterized by Bauer's model.

3. Calculation of Velocity Distribution as a Function of z

Figure 1 presents the XYZ coordinate system. The force balance of a fluid can be expressed as follows [10]:

$$\frac{\partial \tau_x}{\partial z} = \frac{\partial P}{\partial x} \qquad (1)$$

$$\frac{\partial \tau_y}{\partial z} = \frac{\partial P}{\partial y} \qquad (2)$$

where P is the pressure of the fluid, and τ_x and τ_y are the shear stresses in the X- and Y-directions, respectively. The parameters τ and $\dot{\gamma}$ are defined as follows:

$$\tau = \sqrt{\tau_x^2 + \tau_y^2} \qquad (3)$$

$$\dot{\gamma} = \sqrt{\left(\frac{\partial u}{\partial z}\right)^2 + \left(\frac{\partial v}{\partial z}\right)^2} \qquad (4)$$

In Bauer's model, τ is assumed to be represented as a function of $\dot{\gamma}$, as follows [1,14,15]:

$$\tau = \left(\tau_0 + k_1 \cdot \dot{\gamma} + k_2 \cdot \dot{\gamma}^n\right) \cdot \frac{\eta_n}{\eta_0} \quad (5)$$

Here, τ_0, k_1, k_2, and n are Bauer's rheological parameters, and η_n and η_0 represent the P dependent viscosity and ambient viscosity of the base oil, respectively. In this work, according to Dong and Qian [4], the parameters τ_0, k_1, k_2, and n were assumed to be P-independent known values determined from the τ-$\dot{\gamma}$ curve measured at the ambient pressure. In Eyring's model, according to Conry et al. [3] and Johnson and Tevaarwerk [16], the relationship between τ and $\dot{\gamma}$ can be expressed as follows:

$$\dot{\gamma} = \frac{\tau_0'}{\eta_n} \sinh\left(\frac{\tau}{\tau_0'}\right)$$

Here, τ_0' is Eyring's rheological parameter. Therefore, the relationship between τ and $\dot{\gamma}$ can be rewritten as follows:

$$\tau = \tau_0' \cdot \sinh^{-1}\left(\frac{\eta_n \dot{\gamma}}{\tau_0'}\right)$$

The effective viscosity η^* can be defined as follows:

$$\eta^* = \frac{\tau}{\dot{\gamma}} \quad (6)$$

The effective viscosity η^* is a function of $\dot{\gamma}$, which in turn is a function of z. The shear stresses τ_x and τ_y are assumed to be represented as

$$\tau_x = \eta^* \frac{\partial u}{\partial z} \quad (7)$$

$$\tau_y = \eta^* \frac{\partial v}{\partial z} \quad (8)$$

Substituting Equations (7) and (8) into Equations (1) and (2), respectively, yields

$$\frac{\partial}{\partial z}\left(\eta^* \frac{\partial u}{\partial z}\right) = \frac{\partial P}{\partial x} \quad (9)$$

$$\frac{\partial}{\partial z}\left(\eta^* \frac{\partial v}{\partial z}\right) = \frac{\partial P}{\partial y} \quad (10)$$

The pressure gradient vector $\boldsymbol{d}$ is defined as follows:

$$\boldsymbol{d} = \left(\frac{\partial P}{\partial x}, \frac{\partial P}{\partial y}\right) \quad (11)$$

Similar to the method employed by Yang et al. [11], this method involves the flow being divided into Couette and Poiseuille flows. As shown in Figure 2, the Xc-direction is considered to be along $\boldsymbol{d}$, and its direction vector is $\boldsymbol{n}_1$. The Yc-direction is perpendicular to $\boldsymbol{d}$, and its direction vector is $\boldsymbol{n}_2$. The local coordinate system, Oc, Xc, and Yc, at a given point depends on the pressure gradient and is determined simultaneously in the process to obtain the pressure distribution. In the Xc and Yc coordinate system, Equations (9) and (10) can be rewritten as follows:

$$\frac{\partial}{\partial z}\left(\eta^* \frac{\partial u_c}{\partial z}\right) = d \quad (12)$$

$$\frac{\partial}{\partial z}\left(\eta^* \frac{\partial v_c}{\partial z}\right) = 0 \quad (13)$$

$$d = \sqrt{\mathbf{d} \cdot \mathbf{d}} \tag{14}$$

Furthermore, the velocities $u_c(z)$ and $v_c(z)$ satisfy the following boundary conditions:

$$u_c(0) = U^-, \ u_c(h) = U^+, \ v_c(0) = V^-, \ v_c(h) = V^+ \tag{15}$$

Here, U^+ and U^- denote the velocities of the upper and lower surfaces in the Xc-direction, respectively; V^+ and V^- denote the velocities of the upper and lower surfaces in the Yc-direction, respectively. Although the velocities $u_c(z)$ and $v_c(z)$ cannot be represented by polynomials exactly [3], here they are approximated and expanded using polynomials of z so that the velocities satisfy Equation (15), as follows. In this case, the variables a_1 and a_2 are unknown.

$$u_c(z) = \frac{\Delta U}{h} \cdot z + U^- - a_1 \cdot z(h - z) - a_2 \cdot z^2 (h - z)^2 \tag{16}$$

$$v_c(z) = \frac{\Delta V}{h} \cdot z + V^- \tag{17}$$

Here, h is the fluid film thickness, and ΔU and ΔV denote the velocity differences; specifically, $\Delta U = U^+ - U^-$ and $\Delta V = V^+ - V^-$. A higher order term of z, for example, $z^3(h-z)^3$, can also be considered; however, in this work, the lower-order approximation was chosen. As the pressure gradient toward the Yc direction is zero, v_c was assumed to be a Couette flow and approximated considering a linear equation of z. Furthermore, as the pressure gradient toward the Xc-direction is generally non-zero, u_c was assumed to be a Poiseuille flow and approximated using a 4th-order polynomial of z. If $d = 0$, then u_c is also a Couette flow, and Equation (16) can be replaced with the following equation:

$$u_c(z) = \frac{\Delta U}{h} \cdot z + U^- \tag{18}$$

The following equations are derived from Equations (16) and (17):

$$\frac{\partial u_c}{\partial z} = u'(z) = \frac{\Delta U}{h} - a_1 \cdot (h - 2z) - a_2 \cdot 2z(h - z)(h - 2z) \tag{19}$$

$$\frac{\partial v_c}{\partial z} = v'(z) = \frac{\Delta V}{h} \tag{20}$$

The following equations are derived from Equation (19):

$$u'(0) = \frac{\Delta U}{h} - a_1 h \tag{21}$$

$$u'(h) = \frac{\Delta U}{h} + a_1 h \tag{22}$$

$$u'\left(\frac{h}{4}\right) = \frac{\Delta U}{h} - a_1 \frac{h}{2} - a_2 \frac{3h^3}{16} \tag{23}$$

$$u'\left(\frac{3h}{4}\right) = \frac{\Delta U}{h} + a_1 \frac{h}{2} + a_2 \frac{3h^3}{16} \tag{24}$$

If rheological constitutive equations are given, the effective viscosity $\eta^*(z)$ can be calculated using Equations (4)–(6), (16) and (17). The integration of Equation (12) from $z = 0$ to $z = h$ yields Equation (25), and the integration of Equation (12) from $z = h/4$ to $z = 3h/4$ yields Equation (26). These equations are used to determine the values of a_1 and a_2.

$$\eta^*(h) \cdot \left(\frac{\Delta U}{h} + a_1 h\right) - \eta^*(0) \cdot \left(\frac{\Delta U}{h} - a_1 h\right) = dh \tag{25}$$

$$\eta^*\left(\frac{3h}{4}\right) \cdot \left(\frac{\Delta U}{h} + a_1\frac{h}{2} + a_2\frac{3h^3}{16}\right) - \eta^*\left(\frac{h}{4}\right) \cdot \left(\frac{\Delta U}{h} - a_1\frac{h}{2} - a_2\frac{3h^3}{16}\right) = \frac{dh}{2} \quad (26)$$

Equations (21) and (22) show that both $\eta^*(h)$ and $\eta^*(0)$ do not contain a_2. Therefore, Equation (25) does not contain a_2 and contains only the unknown variable a_1. The equation can be solved using the one-variable Newton–Raphson method. Although Equation (26) contains both a_1 and a_2, a_1 has been determined using Equation (25). Consequently, Equation (26) can be considered as an equation involving only the unknown variable a_2. Thus, it can also be solved using the one-variable Newton–Raphson method. To determine a_1, a non-dimensional variable b_1 defined using Equation (27) and a function $f_1(b_1)$ defined using Equation (28) are introduced. The value of b_1 can be calculated considering $f_1(b_1) = 0$.

$$b_1 = \log\left(\frac{2a_1\eta_n}{d}\right) \quad (27)$$

$$f_1(b_1) = \eta^*(h) \cdot \left(\frac{\Delta U}{h} + e^{b_1}\frac{dh}{2\eta_n}\right) - \eta^*(0) \cdot \left(\frac{\Delta U}{h} - e^{b_1}\frac{dh}{2\eta_n}\right) - d \cdot h \quad (28)$$

When b_1 is near the solution, Δb_1 can be calculated using the following equation:

$$0 = f_1(b_1 + \Delta b_1) \fallingdotseq f_1(b_1) + \frac{df_1}{db_1} \cdot \Delta b_1 \fallingdotseq f_1(b_1) + e^{b_1}\frac{dh}{2\eta_n}[\eta^*(h) + \eta^*(0)] \cdot \Delta b_1 \quad (29)$$

In other words, the new candidate $b_{1\text{new}}$ of b_1 is calculated using the iterative process of Newton–Raphson's method, as follows:

$$b_{1\text{new}} = b_1 + \Delta b_1 = b_1 - \frac{f_1(b_1)}{a_1 h [\eta^*(h) + \eta^*(0)]} \quad (30)$$

As $\eta^*(h)$ and $\eta^*(0)$ are originally functions of b_1, df_1/db_1 includes $d\eta^*/db_1$; however, in this work, the dependency was ignored, and Δb_1 was approximated as in Equation (30). To determine a_2, a non-dimensional variable b_2 defined using Equation (31) and a function $f_2(b_2)$ defined using Equation (32) are introduced. The value of b_2 can be calculated considering $f_2(b_2) = 0$.

$$b_2 = \frac{2a_2 h^2 \eta_n}{5d} \quad (31)$$

$$f_2(b_2) = \eta^*\left(\frac{3h}{4}\right) \cdot \left(\frac{\Delta U}{h} + a_1\frac{h}{2} + a_2\frac{3h^3}{16}\right) - \eta^*\left(\frac{h}{4}\right) \cdot \left(\frac{\Delta U}{h} - a_1\frac{h}{2} - a_2\frac{3h^3}{16}\right) - \frac{dh}{2} \quad (32)$$

When b_2 is near the solution, Δb_2 can be calculated using the following equation:

$$0 = f_2(b_2 + \Delta b_2) \fallingdotseq f_2(b_2) + \frac{df_2}{db_2} \cdot \Delta b_2 \fallingdotseq f_2(b_2) + \frac{15dh}{32\eta_n}\left[\eta^*\left(\frac{3h}{4}\right) + \eta^*\left(\frac{h}{4}\right)\right] \cdot \Delta b_2 \quad (33)$$

In other words, the new candidate $b_{2\text{new}}$ of b_2 is calculated using the iterative process of Newton–Raphson's method, as follows:

$$b_{2\text{new}} = b_2 + \Delta b_2 = b_2 - \frac{f_2(b_2)}{[\eta^*(3h/4) + \eta^*(h/4)]} \cdot \frac{32\eta_n}{15dh} \quad (34)$$

As $\eta^*(3h/4)$ and $\eta^*(h/4)$ are originally functions of b_2, df_2/db_2 includes $d\eta^*/db_2$; however, in this work, the dependency was ignored, and Δb_2 was approximated as in Equation (34). Subsequently, in the iteration process of Newton–Raphson's method, only η^* depends on the rheological characteristics. Therefore, if the rheological equation corresponding to Equation (5) is incorporated, any isothermal, non-Newtonian EHL calculation can be performed. As per the Newton–Raphson method, the initial value for both b_1 and b_2 can be zero. As both variables a_1 and a_2 are solved using the one-variable Newton–Raphson method, the calculation can be performed within a reasonable time.

4. Calculation of Equivalent Viscosity, Flow, and Surface Force

The flow q_1 along the n_1 direction can be defined using Equation (16), as follows. The density ρ is assumed to be independent of z.

$$q_1 = \int_0^h \rho u_c \, dz = \int_0^h \rho \left[\frac{\Delta U}{h} \cdot z + U^- - a_1 \cdot z(h-z) - a_2 \cdot z^2(h-z)^2 \right] dz \\ = \rho U h - \frac{\rho a_1 h^3}{6} - \frac{\rho a_2 h^5}{30} \tag{35}$$

Here, U is the average velocity in the Xc-direction and can be expressed as follows:

$$U = \frac{U^+ + U^-}{2} \tag{36}$$

The equivalent viscosity η_{eq} is defined as follows:

$$\eta_{eq} = \frac{5d}{2(5a_1 + a_2 h^2)} \tag{37}$$

Consequently, q_1 can be represented as

$$q_1 = \rho U h - \frac{\rho h^3}{12\eta_{eq}} \cdot d \tag{38}$$

The flow q_2 along the n_2 direction can be derived from Equation (17), as follows:

$$q_2 = \int_0^h \rho v_c \, dz = \rho V h \tag{39}$$

Here, V is the average velocity in the Yc direction and can be expressed as follows:

$$V = \frac{V^+ + V^-}{2} \tag{40}$$

Hence, in the XYZ coordinate system, the flow vector q can be expressed as

$$q = \left(\rho U h - \frac{\rho h^3}{12\eta_{eq}} \cdot d \right) \cdot n_1 + \rho V h \cdot n_2 \\ = \rho U h \cdot n_1 + \rho V h \cdot n_2 - \frac{\rho h^3}{12\eta_{eq}} \cdot d \cdot n_1 \\ = \rho U h - \frac{\rho h^3}{12\eta_{eq}} \cdot \left(\frac{\partial P}{\partial x}, \frac{\partial P}{\partial y} \right) \tag{41}$$

Here, U is the average velocity vector of the upper and lower surfaces, defined as follows:

$$U = U \cdot n_1 + V \cdot n_2 = (U_x, U_y) \tag{42}$$

U_x and U_y denote the average velocities in the upper and lower surfaces in the XY-direction, respectively. When the mass conservation law is applied to Equation (41), the following modified Reynolds equation is obtained.

$$\frac{\partial \rho U_x h}{\partial x} + \frac{\partial \rho U_y h}{\partial y} - \frac{\partial}{\partial x}\left(\frac{\rho h^3}{12\eta_{eq}} \cdot \frac{\partial P}{\partial x} \right) - \frac{\partial}{\partial y}\left(\frac{\rho h^3}{12\eta_{eq}} \cdot \frac{\partial P}{\partial y} \right) = 0 \tag{43}$$

The difference in the representation of Equations (41) and (43) and that of Newtonian flow only pertains to the viscosities η_{eq} and η_n, respectively. In fact, the equivalent viscosity η_{eq} defined by Equation (37) was determined so that Equations (41) and (43) maintain the same form as that of Newtonian flow. Therefore, the EHL calculation procedure for Newtonian flows, such as the method proposed by Venner and Lubrecht [13], can be applied to the current calculation by simply replacing η_n with η_{eq}. The shear stress τ_1 along the Xc-direction can be expressed as

$$\tau_1(z) = \eta^* \frac{\partial u_c}{\partial z} = \eta^* \left[\frac{\Delta U}{h} - a_1 \cdot (h - 2z) - a_2 \cdot 2z(h-z)(h-2z) \right] \quad (44)$$

Therefore, the surface forces P_{10} and P_{1h} acting on the lower and upper surfaces along the Xc-direction, respectively, can be defined as

$$P_{10} = \tau_1(0) = \eta^*(0) \cdot \left(\frac{\Delta U}{h} - a_1 h \right) \quad (45)$$

$$P_{1h} = -\tau_1(h) = -\eta^*(h) \cdot \left(\frac{\Delta U}{h} + a_1 h \right) \quad (46)$$

The shear stress τ_2 along the Yc-direction is expressed as

$$\tau_2 = \eta^* \frac{\partial v_c}{\partial z} = \eta^* \frac{\Delta V}{h} \quad (47)$$

Therefore, the surface forces P_{20} and P_{2h} acting on the lower and upper surfaces along the Yc-direction, respectively, can be defined as

$$P_{20} = \tau_2(0) = \eta^*(0) \cdot \frac{\Delta V}{h} \quad (48)$$

$$P_{2h} = -\tau_2(h) = -\eta^*(h) \cdot \frac{\Delta V}{h} \quad (49)$$

In the XYZ coordinate system, the surface force vectors $\boldsymbol{P}_0$ and $\boldsymbol{P}_h$ that act on the lower and upper surfaces, respectively, can be expressed as

$$\boldsymbol{P}_0 = P_{10} \cdot \boldsymbol{n}_1 + P_{20} \cdot \boldsymbol{n}_2 \quad (50)$$

$$\boldsymbol{P}_h = P_{1h} \cdot \boldsymbol{n}_1 + P_{2h} \cdot \boldsymbol{n}_2 \quad (51)$$

5. Application to a Grease

As mentioned previously, Kochi et al. [1] conducted experiments on grease under soft EHL conditions and measured the film thickness and traction forces. In the present study, the proposed method was applied to one of the greases considered in the study by Kochi et al. [1] so as to validate the theoretical approach. Grease A in the literature [1] was chosen to test the proposed method. The modified Reynolds equation, as expressed in Equation (43), was solved using a multi-level method, as reported by Venner and Lubrecht [13]. The commercial program Tribology Engineering Dynamics Contact Problem Analyzer (TED/CPA) V852 was employed. Figure 3 illustrates the calculation conditions. The upper body was a steel ball, and the lower body was a disk composed of glass or polycarbonate (PC). The rheological properties of the grease were assumed to be represented by Bauer's model, according to existing literature [1]. Detailed properties of the steel, glass, PC, and grease are described in the previous study [1]. The pressure dependency of the density was defined using Dowson–Higginson's formula, as follows:

$$\rho(P) = \rho_0 \cdot \frac{p_0 + \beta \cdot P}{p_0 + P}, \quad \rho_0 = 1, \quad p_0 = 590 \text{ MPa}, \quad \beta = 1.34 \quad (52)$$

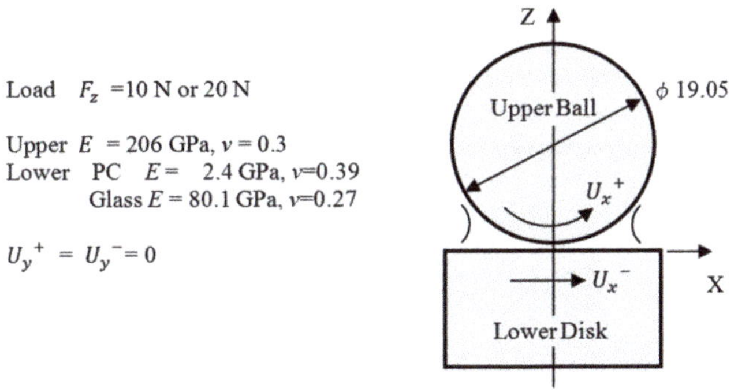

Load F_z =10 N or 20 N

Upper E = 206 GPa, v = 0.3
Lower PC E = 2.4 GPa, v=0.39
        Glass E = 80.1 GPa, v=0.27

$U_y^+ = U_y^- = 0$

Figure 3. Calculation condition.

The pressure dependency of the base oil viscosity was assumed to be defined using Barus' formula:

$$\eta_n(P) = \eta_0 \cdot \exp(\alpha P), \quad \eta_0 : 49.5 \text{ mPa·s}, \quad \alpha : 14 \text{ GPa}^{-1} \tag{53}$$

In particular, Equation (6) diverges when $\dot{\gamma}$ approaches zero. It was assumed that if $\dot{\gamma}$ is lower than a certain value c_{min}=100.0 s^{-1}, then η^* varies linearly with the gradient $d\eta^*/d\dot{\gamma}$ at c_{min}. Bauer's parameter k_1 was assumed to be the base oil ambient viscosity η_0. The values of Bauer's parameters τ_0, k_2, and n were determined from the apparent viscosity curve of grease A, as shown in Figure 9 of Kochi et al. [1]. When $P = 0$, the curve was assumed to pass through the following three points: 100 mm/s, 6.46475 Pa·s; 10,000 mm/s, 0.16712 Pa·s; and 1,000,000 mm/s, 0.06573 Pa·s.

The values of τ_0, k_2, and n were determined to ensure that Bauer's curve passes through the abovementioned three points, as follows:

$$\tau_0 = 0.000621839 \text{ MPa}, \quad k_2 = 6.99118 \cdot 10^{-7}, \quad n = 0.7248 \tag{54}$$

Figure 4 presents the central film thickness of grease A and base oil as a function of the rolling velocity in the case of pure rolling and $F_z = 10$ N. The solid lines show the calculation results and the dotted lines show the experimental results. The experimental data were read using the caliper from Figure 4 of Kochi et al. [1]. Figure 4a,b shows the cases of a PC disk and glass disk, respectively. The calculation range was set as follows: $-1.0 \leq X \leq 0.4$ and $-0.6 \leq Y \leq 0.6$.

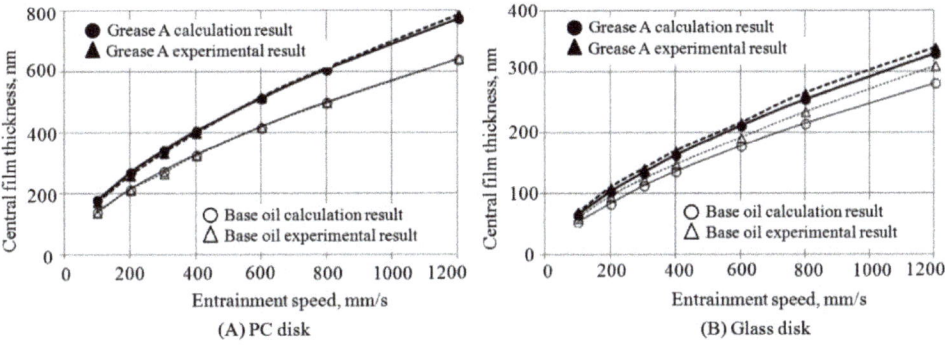

Figure 4. Comparison of film thickness between the calculation and the experiment.

However, in the case of grease A, a glass disk, and a velocity of less than or equal to 300 mm/s, the range was set as follows: $-0.35 \leq X \leq 0.14$ and $-0.2 \leq Y \leq 0.2$.

In these cases, the oil film was thin and the wide range calculation became hard to perform. In the case of the PC disk, the calculation results exhibited good agreement with the experimental results. In the case of the glass disk, the results of the base oil showed some difference but the other data showed reasonable agreement. Figure 5, which illustrates a sample calculation, shows the distribution of P, h, and η_{eq} for the case involving a PC disk, a pure rolling velocity of 1200 mm/s, and grease A. Typically, in the case of pure rolling velocity, $\dot{\gamma}$ is small and η_{eq} is large. In addition, only the appearance of the distribution of η_{eq} differs from that of the base oil. Figure 5d shows the distribution of η_{eq} at section Y = 0. It can be seen that η_{eq} becomes extremely large at the center, where the pressure gradient is small and the flow volume is low.

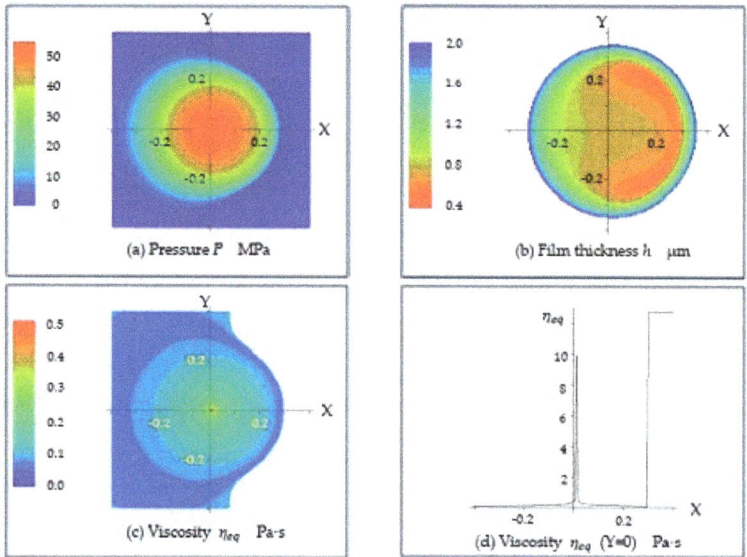

Figure 5. Distribution of P, h, and η_{eq} of grease A at a rolling velocity of 1200 mm/s.

Figure 6 presents the traction coefficient as a function of the slide roll ratio when $F_z = 20$ N. The solid lines show the calculation results and the dotted lines show the experimental results. The experimental data were read using the caliper from Figure 8 of Kochi et al. [1]. The slide roll ratio is the difference between the upper and lower velocities divided by their average value. The traction coefficient was calculated according to the approaches proposed in the existing literature [1,17]:

$$TRC = \frac{TX0 - TXh}{2F_z} \quad (55)$$

Here, $TX0$ and TXh denote the X-direction traction forces acting on the lower and upper surfaces, respectively; F_z is the load. The calculation range was $-0.8 \leq X \leq 0.5$, $-0.5 \leq Y \leq 0.5$. The calculation results were in fairly good agreement with the experimental results.

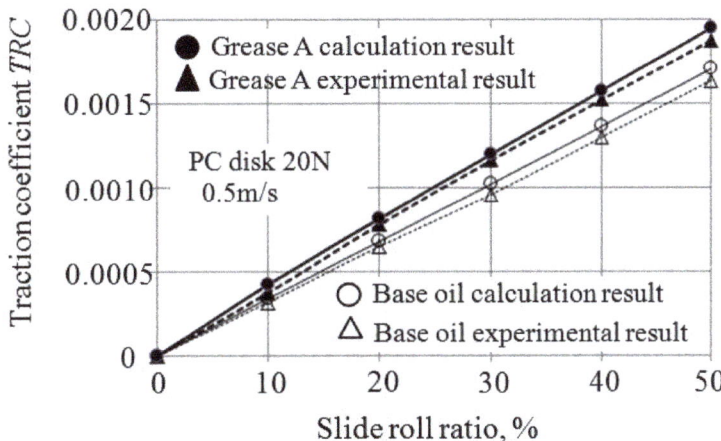

Figure 6. Comparison of the traction coefficient between the calculation and the experiment.

Figure 7, which illustrates a sample calculation, shows the distribution of P, h, and η_{eq} for the case involving grease A and a slide roll ratio of 10%. Only the appearance of the distribution of η_{eq} differs from that of the base oil, exhibiting a figure eight in the vicinity of the contact point. Figure 7c,d shows the same distribution of η_{eq} in different display ranges. It can be observed that η_{eq} reduces in the rapid flow region. The figure eight phenomenon is characteristic of non-Newtonian flow, in which the apparent viscosity becomes large at a low velocity gradient. This phenomenon can be explained as follows. Let the XY coordinates of points A, B, C, and D be $(-0.1, 0)$, $(+0.1, 0)$, $(0, -0.1)$, and $(0, +0.1)$, respectively. The velocity gradient vector n_1 at these points are approximately $(+1, 0)$, $(-1, 0)$, $(0, +1)$, and $(0, -1)$, respectively, as shown in Figure 8a. When a_2 in Equation (37) is neglected, η_{eq} can be expressed as follows:

$$\eta_{eq} = \frac{d}{2a_1} \quad (56)$$

Calculating a_1 from Equation (25) and substituting it into Equation (56) yields:

$$\eta_{eq} = \frac{[\eta^*(h) + \eta^*(0)]}{2} \cdot \frac{1}{1 + \frac{\Delta U}{dh^2}[\eta^*(0) - \eta^*(h)]} \quad (57)$$

At points C and D, the direction of flow and that of n_1 is orthogonal, so $\Delta U = 0$. Consequently, the equivalent viscosity parameters $\eta_{eq,C}$ and $\eta_{eq,D}$ at points C and D are given as follows:

$$\eta_{eq,C} = \eta_{eq,D} = \frac{[\eta^*(h) + \eta^*(0)]}{2} \quad (58)$$

At point A, where n_1 directs towards $+X$ and the velocity gradient at $Z = h$ is greater than that at $Z = 0$ (as shown in Figure 8b), the following equation is satisfied:

$$\eta^*(0) > \eta^*(h), \quad \Delta U = U_x^+ - U_x^- > 0 \quad (59)$$

At point B, where n_1 directs towards $-X$, and the velocity gradient at $Z = h$ is smaller than that at $Z = 0$ (as shown in Figure 8c), the following equation is satisfied:

$$\eta^*(0) < \eta^*(h), \quad \Delta U = -(U_x^+ - U_x^-) < 0 \quad (60)$$

In any case, the equivalent viscosity parameters $\eta_{eq,A}$ and $\eta_{eq,B}$ at points A and B become smaller than $\eta_{eq,C}$ and $\eta_{eq,D}$ by the effect of the second term of Equation (57). In the pure rolling case, where ΔU is 0, Equation (57) results in Equation (58). It can be understood that in such a case, no figure-eight-shaped distribution appears as is shown in Figure 5c.

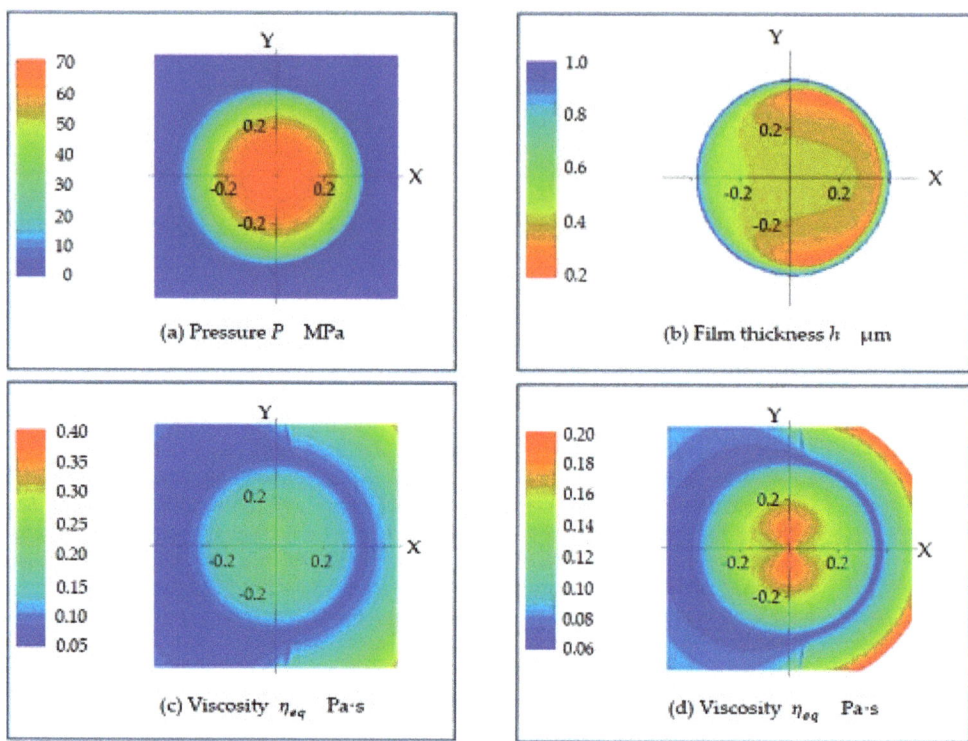

Figure 7. Calculation results of grease A at a 10% slide ratio.

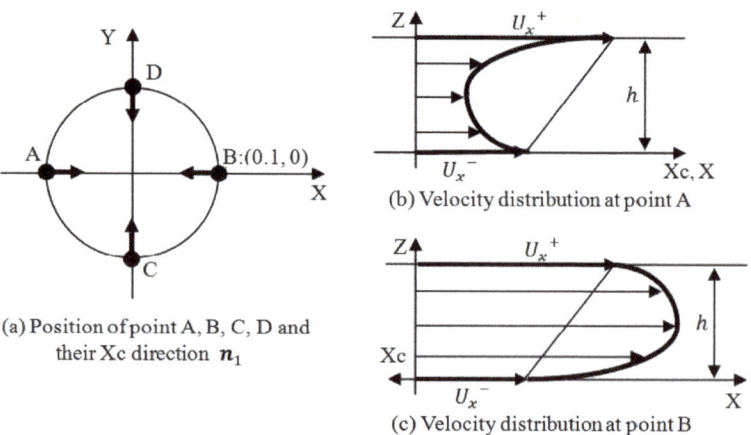

Figure 8. Explanation of the figure eight phenomenon.

6. Conclusions

In this study, an isothermal, non-Newtonian EHL formulation of Bauer's model was performed using the local coordinate system of the pressure gradient. The flow toward the pressure gradient was assumed to be a Poiseuille flow and was approximated using a 4th-order polynomial of z. The flow along the direction of the zero pressure gradient was assumed to be a Couette flow and was approximated using a linear function of z. The following results were obtained.

(1) A modified Reynolds equation, which contains an equivalent viscosity, was obtained.
(2) The EHL calculation procedure for Newtonian flows can be applied to non-Newtonian flows by simply replacing the Newtonian viscosity with the equivalent viscosity.
(3) If rheological equations are incorporated, any isothermal, non-Newtonian EHL calculation can be performed easily.
(4) As the equivalent viscosity is calculated using the one-variable Newton–Raphson method, the EHL calculation can be performed within a reasonable calculation time.
(5) Using Bauer's model, the formulation was applied to a grease that was evaluated experimentally by Kochi et al. [1]. The results obtained using the proposed method and the experimental results were compared, and reasonable agreement was noted.
(6) In the case of sliding velocity, the equivalent viscosity shows a figure-eight-shaped distribution in the vicinity of the contact point.

However, the proposed method yields an approximate solution. If the Poiseuille flow and Couette flow cannot be approximated using a 4th-order polynomial of z and a linear function of z, respectively, the obtained results may be inaccurate. The application limits of the current formulation are not clear. Therefore, future work should be focused on determining these limitations.

Funding: This research was funded by the TriboLogics Corporation.

Acknowledgments: The author sincerely thanks TriboLogics Corporation (www.tribology.co.jp/indexEng.htm, accessed on 13 May 2021) for permitting the use of TED/CPA. The author is also very grateful to Hiroyuki OHTA of the Nagaoka University of Technology, Japan, for his constant guidance, encouragement, and valuable suggestions.

Conflicts of Interest: The author declares no conflict of interest.

References

1. Kochi, T.; Ichimura, R.; Yoshihara, M.; Dong, D.; Kimura, Y. Film Thickness and Traction in Soft EHL with Grease. *Tribol. Online* **2017**, *12*, 171–176. [CrossRef]
2. Kauzlarich, J.J.; Greenwood, J.A. Elastohydrodynamic Lubrication with Herschel-Bulkley Model Greases. *ASLE Trans.* **1972**, *15*, 269–277. [CrossRef]
3. Conry, T.F.; Wang, S.; Cusano, C. A Reynolds-Eyring Equation for Elastohydrodynamic Lubrication in Line Contacts. *J. Tribol.* **1987**, *109*, 648–654. [CrossRef]
4. Dong, D.; Qian, X. A theory of elastohydrodynamic grease-lubricated line contact based on a refined rheological model. *Tribol. Int.* **1988**, *21*, 261–267. [CrossRef]
5. Peiran, Y.; Shizhu, W. A Generalized Reynolds Equation for Non-Newtonian Thermal Elastohydrodynamic Lubrication. *J. Tribol.* **1990**, *112*, 631–636. [CrossRef]
6. Kim, K.H.; Sadeghi, F. Non-Newtonian Elastohydrodynamic Lubrication of Point Contact. *J. Tribol.* **1991**, *113*, 703–711. [CrossRef]
7. Ehret, P.; Dowson, D.; Taylor, C.M. On lubricant transport conditions in elastohydrodynamic conjuctions. *Proc. R. Soc. A Math. Phys. Eng. Sci.* **1998**, *454*, 763–787. [CrossRef]
8. Greenwood, J.A. Discussion: Two-dimensional flow of a non-Newtonian lubricant. *Proc. Inst. Mech. Eng. Part J J. Eng. Tribol.* **2001**, *215*, 121–122. [CrossRef]
9. Sharif, K.J.; Kong, S.; Evans, H.P.; Snidle, R.W. Contact and elastohydrodynamic analysis of worm gears Part 1: Theoretical formulation. *Proc. Inst. Mech. Eng. Part C J. Mech. Eng. Sci.* **2001**, *215*, 817–830. [CrossRef]
10. Liu, X.; Jiang, M.; Yang, P.; Kaneta, M. Non-Newtonian Thermal Analyses of Point EHL Contacts Using the Eyring Model. *J. Tribol.* **2005**, *127*, 70–81. [CrossRef]
11. Yang, Q.; Huang, P.; Fang, Y. A novel Reynolds equation of non-Newtonian fluid for lubrication simulation. *Tribol. Int.* **2016**, *94*, 458–463. [CrossRef]
12. Bordenet, L.; Dalmaz, G.; Chaomleffel, J.-P.; Vergne, F. A study of grease film thicknesses in elastorheodynamic rolling point contacts. *Lubr. Sci.* **1990**, *2*, 273–284. [CrossRef]
13. Venner, C.; Lubrecht, A. Foreword. *Veh. Tribol.* **2000**, *37*, 7–8. [CrossRef]
14. Bauer, W.H.; Finkelstein, A.P.; Wiberley, S.E. Flow Properties of Lithium Stearate-Oil Model Greases as Functions of Soap Concentration and Temperature. *ASLE Trans.* **1960**, *3*, 215–224. [CrossRef]

15. Palacios, J.; Palacios, M. Rheological properties of greases in ehd contacts. *Tribol. Int.* **1984**, *17*, 167–171. [CrossRef]
16. Johnson, K.L.; Tevaarwerk, J.L. Shear behaviour of elastohydrodynamic oil films. *Proc. R. Soc. Lond. Ser. A Math. Phys. Sci.* **1977**, *356*, 215–236. [CrossRef]
17. De Vicente, J.; Stokes, J.R.; Spikes, H.A. Rolling and sliding friction in compliant, lubricated contact. *Proc. Inst. Mech. Eng. Part J J. Eng. Tribol.* **2006**, *220*, 55–63. [CrossRef]

Perspective

Grease Performance Requirements and Future Perspectives for Electric and Hybrid Vehicle Applications

Raj Shah [1,*], Simon Tung [2], Rui Chen [3,*] and Roger Miller [4]

1. Koehler Instrument Company, Inc., 85 Corporate Drive, Holtsville, NY 11742, USA
2. Consultant, Innovation Technology Consulting Inc., Roseville, MI 48066, USA; simontung168@gmail.com
3. Department of Materials Science and Chemical Engineering, Stony Brook University, Stony Brook, NY 11794, USA
4. Chemical Engineer, 18342 Char A Banc, Baton Rouge, LA 70817, USA; geauxtigers1@msn.com
* Correspondence: rshah@koehlerinstrument.com (R.S.); rui.chen.2@stonybrook.edu (R.C.)

Abstract: Electric vehicle sales are growing globally in response to the move towards a greener environment and a reduction in greenhouse gas emissions. As in any machine, grease lubricants will play a significant role in the component life of these new power plants and drivetrains. In this paper, the role of grease lubrication in electric vehicles (EVs) and hybrid vehicles (HVs) will be discussed in terms of performance requirements. Comparisons of grease lubrication in EVs and HVs for performance requirements to current internal combustion engines (ICEs) will be reviewed to contrast the major differences under different operating conditions. The operating conditions for grease lubrication in these EVs and HVs are demanding. Greases formulated and manufactured to meet specific performance specifications in EVs and HVs, which will operate within these specific electrification components, will be reviewed. Specifically, the thermal and electrified effects from the higher operating temperatures and electromagnetic fields on lubricant degradation, rheology, elastomer compatibility, and corrosion protection of the grease need to be evaluated to accurately meet the performance requirements for EVs and HV. The major differences between EVs and conventional ICEVs can be grouped into the following technical areas: energy efficiency, noise, vibration, and harshness (NVH), the presence of electrical current and electromagnetic fields from electric modules, sensors and circuits, and bearing lubrication. Additional considerations include thermal heat transfer, seals, corrosion protection, and materials' compatibility. The authors will review the future development trends of EVs/HVs on driveline lubrication and thermal management requirements. The future development of electric vehicles will globally influence the selection and development of gear oils, coolants, and greases as they will be in contact with electric modules, sensors, and circuits and will be affected by electrical current and electromagnetic fields. The increasing presence of electrical parts in EVs/HVs will demand the corrosion protection of bearings and other remaining mechanical components. Thus, it is imperative that specialized greases should be explored for specific applications in EVs/HVs to ensure maximum protection from friction, wear, and corrosion to guarantee the longevity of the operating automobile. Low-viscosity lubricants and greases will be used in EVs to achieve improvements in energy efficiency. However, low-viscosity fluids reduce the film thickness in the driveline application. This reduced film thickness increases the operating temperature and reduces the calculated fatigue life of the bearings. Bearing components for EVs/HVs will be even more crucial as original equipment manufacturers (OEMs) specify these low-viscosity fluids. The application of premium bearing components using low-viscosity grease will leverage materials, bearing geometries, and surface topography to combat the impact of low-viscosity lubricants. In addition, EVs and HVs will create their own NVH challenges. Wind and road noise are more prominent, with no masking noise from the ICE. Increasing comfort, quality, and reliability issues will be more complicated with the introduction of new electrified powertrain and E-driveline subsystems. This paper elaborates on the current development trends and industrial test standard for the specified grease used for electrical/hybrid driveline lubrication.

Citation: Shah, R.; Tung, S.; Chen, R.; Miller, R. Grease Performance Requirements and Future Perspectives for Electric and Hybrid Vehicle Applications. *Lubricants* **2021**, *9*, 40. https://doi.org/10.3390/lubricants9040040

Received: 6 February 2021
Accepted: 26 February 2021
Published: 6 April 2021

Publisher's Note: MDPI stays neutral with regard to jurisdictional claims in published maps and institutional affiliations.

Copyright: © 2021 by the authors. Licensee MDPI, Basel, Switzerland. This article is an open access article distributed under the terms and conditions of the Creative Commons Attribution (CC BY) license (https://creativecommons.org/licenses/by/4.0/).

Keywords: greases; EVs; hybrid vehicle; driveline lubricant; electrification; thermal properties; electromagnetic field; noise, vibration and harshness (NVH); energy efficiency

1. Introduction

Due to the amount of wear and friction present in an automotive vehicle, lubricants are essential for the longevity and function of cars. Greases, which are boundary lubricants, are essential to the protection of car parts, preventing them from breaking down. Not only are they formulated specifically for the purpose of preventing wear, but they are also made to withstand the conditions of a car in action. Greases prevent rusting and accumulation of debris on surfaces by forming a protective layer; in addition, greases' many properties such as their ability to flow at high temperatures while also being excellent at dissipating heat are what make them valuable and widely used [1]. A vast majority of applied greases will last throughout the entire lifetime of vehicles and do not need to be reapplied.

Greases are commonly derived from petroleum or synthetic materials. Synthetics are often preferred as they can typically function over a wider range of temperatures compared to those made from petroleum [2]. Furthermore, different types of vehicles can operate under different conditions and with either a higher or lower number of variables. For example, vehicles operating under higher/lower moisture, extreme loads, and having high speed bearings will all require different grease specifications to properly protect against corrosion and wear. This can be done through the usage of thickeners and additives such as rust inhibitors and anti-wear and friction-reducing agents [3].

In electrical vehicles (EVs), greases need to be formulated for new factors, the major ones being the increased presence of electricity, electrical currents, and noise in an EV due to the absence of an internal combustion engine (ICE) [4]. Due to the increased number of electrical components such as electric modules and sensors, the greases must be formulated to be unreactive with electricity. Usually, the noise of an engine will mask the creaks and rattles of a car, but as an EV is silent, any noise from the lack of lubrication and contact of surfaces will become much more apparent. Furthermore, a larger amount of grease must be used in an EV than in an internal combustion engine vehicle (ICEV) as the need for lubrication is higher in an EV. To further the understanding of the need of greases in EVs, this perspective paper will focus on the new variables that are present in EVs that necessitate a specified formulation of greases.

2. Greases in Automotive

2.1. Types of Greases and Their Usages

The most common types of greases are soaps, of which different options are shown in Table 1 along with their properties and applications. Lithium greases are widely used for their lubricity, shear stability, and thermal resistance [3,5]; calcium-based ones have better water resistance but worse thermal resistance, and sodium greases have high dropping points but cannot operate above 120 ° [6]. These greases are created through the blending of base oils and thickeners through intense mixing until the mixture becomes gelatin-like [3,7]. Overall, lithium grease has been shown to impart the advantages of high adherence, non-corrosiveness, available at high pressures, and moisture resistance, making it compatible with several OEM applications such as EVs/hybrid vehicles (HVs) [8,9]. Other types of greases include polyuria (PU), clay, and silica. PU greases' high operating temperature, oxidative protection properties, and low bleed features make them useful for permanent sealing applications [10]. Clay greases are inert and favored in the food industry [3]. Although silica-based greases are a strong thickener, they are very sensitive to heat.

Table 1. Greases containing mineral oil (soaps) (Adapted from [8]).

Soap Base or Thickener	Min. Drop pt. °C (°F) American Society for Testing and Materials (ASTM) D556 IP 132	Min. Usable Temp. °C (°F)	Max Usable Temp. °C (°F)	Rust Protection	Available with Extreme Pressure (EP) Additive	Use	Cost	Official Specification
Lime (calcium)	90 (190)	−20 (0)	60 (140)		Yes	General purpose	Low	BS 3223
Lime (calcium), heat stable	99.5 (21) or 140 (280)	−20 (0) −55 (−65)	80 (175) *	Sometimes	Yes	General purpose and rolling bearings	Low	BS 3223 DEF STAN 91-17 (LG 280) DEF STAN 91-27 (XG 279) MIL-G-10924B
Sodium, conventional	205 (400)	0 (32)	150–175 (300–350)	Yes	No	Glands, seals low–medium-speed rolling bearings	Low	
Sodium and calcium (mixed)	150 (300)	−40 (−40)	120–150 (250–300)	Yes	No	High-speed rolling bearings	Medium	DEF 2261A (XG 271) MIL-L-7711A
Lithium	175 (350)	−40 (−40)	150 (300)	Yes	Yes	All rolling bearings	Medium	DEF STAN 91-12 (XG 271) DEF STAN 91-28 (XG 274) MIL-L-7711A
Aluminum complex	200 (390)	−40 (−40)	160 (320)	Yes	Mild EP	Rolling bearings	Medium/high	
Lithium complex	235 (450)	−40 (−40)	175 (350)	Yes	Mild EP	Rolling bearings	Medium	
Clay	None	−30 (−20)	205 (400) *	Poor	Yes	Sliding friction	Medium	

* Depending on conditions of service.

Overall, approximately six different types of greases are present in a vehicle system—the majority of which are soap greases. Lithium greases are present in car hinges to gears. White greases are used for water sealing and surface adhesion. Aluminum greases are often used in wheel bearings, and copper greases are used in exhaust assemblies and battery pole connections. Finally, red rubber greases are present in areas with O-rings and hydraulic systems, and brake greases are used in drum and disc brakes [11].

2.2. Load

Viscosity does play a critical role in determining load-carrying capacity in fluid film hydrodynamic bearings. Load-carrying capacity is directly proportional to viscosity and operating speed and inversely proportional to the square of film thickness.

For EHL (ElastoHydrodynamic Lubrication) conditions, due to the nature of the contact of the roller elements and the race—Hertzian by nature—the elastic deflection of mating surfaces significantly influences the load being driven by the material properties of the bearing vs. the viscous-elastic properties of the oil phase in grease. For a Newtonian oil, the following equations govern the film thickness [12,13]:

$$\text{LINE CONTACT:} (h_o/R) = (2.65\ G^{0.54}\ U^{0.7})/W^{0.13} \quad \text{(Dowson, 1970)} \tag{1}$$

$$\text{POINT CONTACT:} (h_o/R_x) = 3.63\ U^{0.68}\ G^{0.49}\ W^{-0.073}\ (1 - e^{-0.68k}) \tag{2}$$

where viscosity is contained within the Speed Parameter (U) and load is contained within the Load Parameter (W). In EHL conditions, the film thickness is minimally affected by the load, as seen by the magnitude of the powers for each W shown in the equations.

2.3. Viscosity

Generally, the selection of viscosity is determined by bearing speed, operating temperature, bearing dimensions, and race surface finish. For roller element bearings, the focus will be on EHL conditions, which covers many of the conditions these bearings operate in.

The previous equations relate the variables needed to arrive at the optimal viscosity by determining film thickness. More specifically, they relate surface speed and required viscosity to determine the minimum EHL film thickness needed to match the composite surface roughness in the loaded contact area. From our previous discussion on load, we will focus on these variables given that the contribution of load to this determination is minimal under EHL conditions. There are graphical relationships that relate viscosity and speed to determine the optimal viscosity.

The next step is to determine the correct viscosity at the bearing operating temperature. Using the readily available viscosity curves, the ISO grade for the application can be determined.

Final confirmation of the correct operating viscosity is determined by calculating the specific film thickness:

$$\text{Lambda} = \text{Specific Film Thickness} = \frac{h}{\sigma} = \lambda$$

where h = film thickness from the previous equations and

σ = measured surface roughness (r.m.a., microns)

The choice of the correct viscosity minimizes asperity contact in the element load zone. Specific Film Thickness confirms the chosen viscosity, estimated from bearing speed and temperature, and yields the film thickness for the specific bearing surface finish characteristics. Lambda can have a significant effect on bearing fatigue life.

Lambda values below 1.5 reflect a dramatic loss of bearing fatigue life. Under this condition, the film thickness is not high enough to cover the average asperity height, leading to excessive friction and wear of the race and roller element surfaces. Lambda values between 1.5 and 3.0 contribute to an increased fatigue life, reflecting the ability of the film to cover load zone asperities. Lambda values above 3.0 exhibit increasing but diminishing improvements in fatigue life.

Determination of the correct viscosity is followed by determination of the National Lubricating Grease Institute (NLGI) grade needed for the application. The next step is determining the temperature range at which the grease will work. Care must be taken so as to not over- or underestimate the operating range, as catastrophic failure can occur. Therefore, it is always advised to measure the temperatures of the components if possible. Oftentimes, multiple greases can be found suitable to a particular application, at which point it should come down to the costs unless the item is already on hand [8].

2.4. Grease Formulation Issues

Since greases are formulated to operate within a specific range of conditions, operating outside those conditions or prolonged exposure to high temperatures will have a negative impact on the longevity of said grease. Figure 1 shows a correlation between induction time and the grease life—data obtained from the ASTM D 3527 test. A relationship is noticeable in that as the induction time increases, so does the life of the grease. Further testing in Figure 2 shows that an increase in temperature also leads to a decrease in induction time, thereby reducing the life of the grease. This is because greases with longer induction times achieve higher oxidative stability [14].

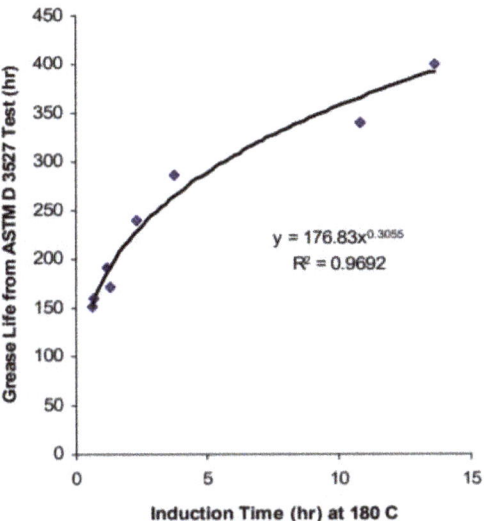

Figure 1. Induction time vs. grease life (ASTM D 3527 Test) [14].

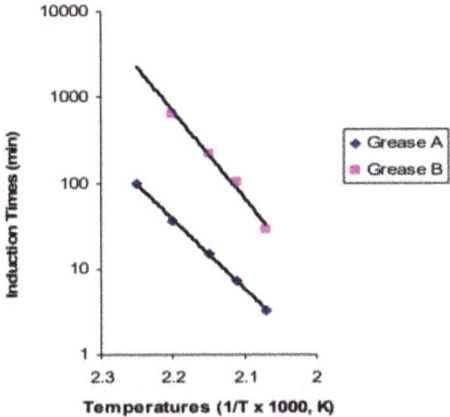

Figure 2. Temperature effects on induction time [14].

3. Performance Requirements for Grease Used in EVs and HVs

In EVs/HVs, a new configuration was designed to use an electrical motor combined with a battery module system for generating energy power. Instead of an engine, which is lubricated with oil and transfers power to a transmission and from there to the wheels, a battery module system powers an electric motor that drives the wheels. When designing EVs, lubrication engineers must select gear oils, coolants, and greases to meet new driveline requirements. The EV/HV configuration design has created substantial impact on driveline lubrication and thermal cooling requirements in the areas of electrical and thermal transfer characteristics, energy efficiency, and the presence of electric currents and magnetic fields. In addition, more system-related requirements such as noise, vibration, and harshness (NVH), seals, and materials' compatibility are also being considered for important performance characteristics.

3.1. Electrical and Thermal Characteristics for Grease Used in EVs and HVs

In many HV or EV advanced system designs, automotive lubricants such as drivetrain fluids or grease lubricating encounter the integrated electric motor (e-motor) and thermal management devices. This leads to the addition of electrical and thermal properties which must be considered on top of conventional lubricant properties. The introduction of electrification components has been targeted for energy efficiency and long-term durability. Automotive industries have asked for the implementation of specialized automotive lubricants or driveline fluids to allow for appropriate thermal cooling specifications, to present bearing protection, to ensure corrosion protection, and to offer oxidation and sludge control.

Recently Tung, Woydt, and Shah published a paper related to the future trends of HV/EV driveline lubrication and thermal management [9]. They also reported that grease specifically designed for driveline lubrication of HVs and EVs should include appropriate electrical properties to guarantee protection from corrosion and be well suited with insulating materials [9,15,16]. There are two additional considerations on electrical conductivity and thermal conductivity of fluids. If the electrical conductivity of the liquids exceeds a certain value, current leaking can occur. However, if it is too low, degradation of the oil can occur because of lubricant oxidation, a direct outcome of electrical arcing in oil, and leads to a decrease in the protective ability of the fluid. Lubricants are not conductive but rather dissipative, though additives can affect the level of electrical conductivity. Conductivity might increase, however, as oxidation causes an oil to deteriorate. Additive suppliers suggested that lubricants susceptible to oxidation are potentially problematic. They also need to have appropriate thermal transfer characteristics, offer high-speed bearing protection, and provide oxidation and sludge control.

3.2. Electric Field Interactions with Driveline Lubricant or Grease Used in EVs/HVs

To understand driveline lubrication under the electric field is very crucial for EV/HV applications by automotive and lubricant suppliers. In a recent publication by Chen, Liang, and co-authors [15], they pointed out that lubricant properties under electric field interaction must be investigated. These important properties can be described as the electrostatic interaction, the electric charge distribution, the formation of transfer film/structural change, and the chemical–physical property changes. It has been found that lubrication is aided by weak electrostatic interactions. Static charges and the transient polarized charges on surfaces, which may be induced and strengthened by the externally applied field, enhance electrostatic interactions. They also claimed that at low electric potential, wear is adhesive-type dominated, while it is abrasive-type dominated when the potential is high. DC has been observed to enhance friction while the friction is reduced by AC. This interaction mechanism is due to vibration induced by the electrostatic force which is fluctuating under electric field. Structural change/oxidative transfer film formation in some material combinations (e.g., graphite–graphite, graphite–copper) has been found responsible for increased wear and reduced friction under the application of the external electric field [9,15].

In addition, Rhee, Yan, He, Xie, and Luo [14–16] indicated that chemical reactions and physical absorption occur at material interfaces under the influence of an external electric field, leading to a change in surface friction and lubrication behavior. Electric carrier (or electron–hole) charge distribution through the formation of localized quantum dots and electron–hole recombination affects interfacial mobility and surface friction properties [14–16].

3.3. Electric Breakdown Mechanisms of Lubricants and Grease

Besides the electrical properties of driveline lubricants or lubricating grease described in the above section, automotive R&D scientists [9,15–18] have reported that a highly fluctuating charged environment requires specially tailored lubricants to avoid component damage and premature failure due to improper lubrication. Major lubrication failure mechanisms explored can be classified into lubricant degradation, microbubble formation, and electrowetting. In the lubricant degradation mechanism, the lubricant base oil and thickeners undergo chemical oxidation to form carboxyl compounds. Lubricity is lost on account of the formation of highly viscous and acidic degradation products and agglomeration of additives. Heat generation causes faster base oil evaporation. Local overheating in EV/HV lubricants can lead to microbubble formation, which may then be driven by viscous drag, pressure gradient, and dielectrophoretic forces. The formation of these microbubbles, which are unstable and coalesce, tends to destabilize lubrication upon electrical breakdown. The microbubbles form more rapidly in conditions of electrode insulation. In addition, the electric field induces interfacial stress on a non-polar lubricant confined between two metallic surfaces in the electrowetting mechanism. Due to differing dielectric properties, a two-phase dispersion of lubricant may also destabilize this mechanism and can lead to the spread and breakdown of the lubricant when the electrostatic stress is too high.

3.4. Thermal Cooling Requirements for Lubricants and Grease Used in Electrical/Hybrid Systems

The heat generation rates in engine power controllers, computer chips, and optical devices/systems are on the rise because of future development trends that favor higher speeds and smaller features for increased performance for engine components, microelectronic devices, and brighter beams for optical devices [9]. Thermal cooling has become one of the main focuses of advanced industries such as microelectronics, transportation, manufacturing, and metrology. Electric hybrid and fuel cell vehicles use power electronics to control their electric motor. Power electronics require their own cooling loop including a heat exchanger, pump, and radiator. Power densities exceeding 100 W/cm^2 while needing to maintain a temperature below 125 °C may eventually exceed 250 W/cm^2. Conventional cooling methods to promote heat rejection rates apply increased surface areas such as fins and microchannels for heat dissipation. However, current thermal cooling designs have already reached their threshold. For HVs or EVs, the cooling requirements are more stringent than IC engines, especially in the case of fast charging and heavy consumption. Lithium-ion batteries and EV motor systems need to maintain the correct temperature range by cooling means. If they exceed this range, the batteries will face a "runaway", not deliver the same power, and, more importantly, they will degrade quickly. Power electronics are also very susceptible to heat, especially during recharging. Heat sinks are used to draw away heat. Solutions for the thermal management of EVs or HVs have been described in the recent publication [9] by Tung, Woydt, and Shah. For example, advanced thermal management approaches such as microchannels for cooling batteries or fast-charging cables or immersion of battery cells in a dielectric fluid have been commercialized in energy and automotive industries.

Thermal Management and Measurement of Thermal Conductivity of Driveline Grease

The future development of thermal management technologies is encumbered by the urgent demand for thermal protection and cooling of electrification components. The traditional method for thermal cooling can no longer be progressed. New requirements have been enforced for high-performance cooling in electrical vehicles, batteries, motors, and power electronics [9]. New developments in thermal management of EVs/HVs are helping to extend the driving range and lifetime requirement. Global research activities using advanced coolants have the scientific merits and high potential for application in thermal cooling technologies. To meet these thermal management requirements, automakers are using combinations of cooling fluids and advanced thermal cooling devices throughout the electrical/hybrid propulsion system to improve overall energy efficiency and coolant compatibility with electrification components. In the next few sections, the authors will review the state-of-the-art thermal management technology used for meeting thermal cooling and extended driving range requirements of EVs/HVs.

Recently, lubricant industrial researchers [18–23] have investigated the thermal properties of lubricants or grease operated under tribological sliding conditions. Pettersson and Callen [18,19] have shown the fundamental phenomenon that the base oil molecular structure determines the thermal capacity and thermal conductivity of a lubricant [18].

The higher the number of the rotational and vibrational quantum states, the higher the thermal capacity [19]. When there are multiple vibrational and rotational states, it takes a higher energy input to increase the averaged kinetic energy, e.g., the temperature. In addition, Gedde and Jin [20,21] have indicated that the thermal conductivity of base oil was correlated to the molecular diffusivity in the fluid. The more easily the molecules of a lubricant pass through each other, the higher the lubricant thermal conductivity. This also means that there is a relationship between the lubricant viscosity and lubricant thermal properties because both the molecular quantum state density and the diffusivity closely correlate to the lubricant viscosity. This correlation can restrict the selection of the lubricant when both the tribological working condition and thermal management are considered. When tribological working conditions take a higher priority, it is difficult to change the base oil thermal properties. Thus, it is quite desirable to change the lubricant thermal property with some additives.

Jin, Shaikh, and Barbés have found out that driveline lubricants can significantly increase the thermal conductivity and thermal capacity of a lubricant by adding nanoparticles to the lubricant [21,22]. Essentially, adding those dispersed nanoparticles increased the carriers of thermal energies. Adding 0.8 vol% of silica nanoparticles can double the thermal conductivity of a lubricant [21]. Polyalphaolefin (PAO) containing 0.5 vol% carbon nanotubes has a more than 50% thermal conductivity compared to neat PAO. However, nanoparticles also lower the specific heat of the lubricant [23]. In addition, Chen and Dai and colleagues also investigated the synergistic effects of nano-lubricant additives [24,25]. They indicated that this nano-lubricant additive can be used to optimize the thermal property of a lubricant to fit any specific powertrain cooling design. Moreover, the nanoparticle additive improves the tribological performance of lubricants. They have shown substantial experimental evidence using this nano-lubricant additive which can dissipate heat extremely fast in EV/HV cooling processes.

In the lubricant industry, the most common experimental method for measuring the thermal conductivity of a lubricant is called the transient hot-wire method [26,27]. The transient hot-wire experimental set-up was simple to perform and had high accuracy. This method used a Pt or Ni wire which was sealed inside a cylindrical pressure vessel filled with lubricant. The wire was heated up for a short amount of time electrically, and its temperature was monitored simultaneously by its electric resistance. The thermal conductivity and the thermal capacity of the lubricant can be calculated from the temperature change of the wire. In general, this measurement can be modeled as an axisymmetric thermal transportation problem [26]. It has an additional advantage when used to characterize lubricants, as the lubricant thermal properties are highly correlated with its pressure, and the pressurized transient hot-wire method is easy to achieve.

3.5. The Other Requirements for Grease Used in EVs and HVs

Another three major issues that are present in EVs or HVs are energy efficiency, noise, and electrification components [28]. Another important aspect of an EV is the increased presence of electrification components and how the greases will be affected by the electromagnetic fields and electrical currents. These place an even higher emphasis on corrosion prevention as the electrical currents will corrode the metallic components at a faster rate as compared to those in a regular ICE. In this perspective paper, we will discuss these important areas and their specific requirements used in EVs/HVs in the following sections.

3.5.1. Energy Efficiency

While EVs are extremely efficient in energy consumption, reports have shown that as much as 57% of the energy used by the car is for overcoming frictional losses [29]. The main components considered to make up the large frictional loss include motor, tires, steering system, wheel bearings, joints, suspensions, and a few others [29]. Friction losses observed in electrical motors (EMs) are attributed to the heat produced by rotors, vibration, wear, and going against wind resistance. In rolling bearings, the main sources of friction loss are from churning, sliding, which is the source of the greatest friction loss of bearings, rolling, and seal sliding friction [30]. For these, a good lubricant or grease is expected to properly offset the frictions. To improve the energy efficiency of grease in these bearing applications, synthetics, primarily polyalphaolefins (PAOs), are used. Given the mix of sliding and rolling contact in the load zone of point contact bearings (ball and spherical roller bearings), PAO-formulated greases have improved traction properties compared to mineral oil-formulated greases. Greases are mainly used in bearings and gears due to them being more of a solid than oil lubricants, as they will not leak out of components and cause damage. Since the usage of greases is much easier with bearings and gears, about 85% of all bearings are lubricated with grease. As increasing energy efficiency is a big focus in the EV industry, all opportunities to reduce friction will help achieve the goal of reaching a range of up to 400 miles [31].

Energy efficiency can be correlated to lubricating film thickness. Thinner lubricants reduce viscous friction, allowing more energy to be conserved. However, greases will reduce lubricating film thickness at high temperatures and bring their own challenges on wear protection of sliding surfaces. A thinner lubricating film will be closer to the mixed and boundary lubrication regimes where wear is a concern. Striking a balance between the capability to remain in a full film lubrication regime and the wear protection towards thinner lubricating films is a crucial turning point that enhances energy efficiency using low-viscosity lubricant or grease. Several papers have emphasized that a new approach using either surface coatings or lower-roughness surfaces will ensure a thin lubricating film to separate surfaces and mitigate wear.

One of the most significant challenges for EVs is to extend the driving range besides energy efficiency. This challenge also imposes the infrastructure of wide-spread recharging stations. Without recharging infrastructure, all passengers must be prepared to give up the 300- to 400-mile ranges they are accustomed to in a conventional vehicle. Automotive engineers are looking for innovation approaches to improve energy efficiency by making vehicles lighter and by reducing torque in all components.

3.5.2. Electrification Components

The projected growth of EVs and HVs in the automotive industries has made evolutional changes in propulsion system electrification components. Powertrain and driveline components will be enhanced by electric motors and E-drive configuration. Electrification components including electric motors or regenerative brake components will be integrated in this advanced configuration with thermal management systems for cooling applications. Electric motors and power electronics will be in contact with the cooling fluids. Coolants are required to cool motors and power electronics by removing heat for thermal management. In addition, automotive lubricants for electrified propulsion systems must function as effective coolants and reduce the corrosion of copper windings, on top of needing composites and rare earth magnetic materials while upholding wear and oxidation protection and energy efficiency.

Dr. Kuldeep K. Mistry [28] at The Timken Co. in Ohio also indicated that the development of electric vehicles will influence the selection and development of gear oils, coolants, and greases, because the lubricating oils or greases will be in contact with electric modules, sensors, and circuits. It will be affected by electrical current and electromagnetic fields. Soon, the number of electrical connections is expected to quadruple. In this context, one of the key life-limiting considerations will be corrosion.

Although grease oils will prevent moisture from reaching the surfaces, thus preventing rust, the presence of electromagnetic fields will create another challenge. Electromagnetic fields can propagate without the need for a solid medium. The amount of energy transferred is dependent on the intensity of the field, the frequency of its oscillation, and the dielectric properties of the material. The more energy the object absorbs, the more quickly the object will be heated. The object will gain heat as the frequency of the electromagnetic field increases. However, if the temperature diffusion rate is slower than the rate at which the electromagnetic field releases its energy, the temperature will increase rapidly [31]. As electric motors produce electromagnetic fields, there is a chance for the premature degradation of greases. The electrical discharge and free radicals will react with the oxygen in greases, creating hydrogen peroxide and continuing the chain of free radical reactions. This leads to the oxidation of base oils and thickeners, which causes a loss in lubricity. In addition, components within the grease will start to separate and the thermal effects from the electrical discharge will cause certain parts to evaporate and eventually cause grease failure [18,31]. To this degree, experiments have shown that a conductive lubricant will help with its corrosion prevention. This is because non-conductive greases will trap the energy from electrical currents and electromagnetic fields, leading to a sudden release and causing great damage [18,31]. However, if the material is conductive, then while the electrical currents will pass through, it will be in a much smaller amount and cause less damage.

3.5.3. Noise

The absence of an ICE to mask vibrations, harshness, squeaks, and rattles will trigger more problems such as NVH and buzz, squeak, and rattle (BSR). Chad Chichester has reported, in a recent Society of Tribologists and Lubricant Engineers (STLE) publication [28,29], that NVH and BSR can affect sensors that are increasingly used in vehicle safety and guidance. The choice of greases used in EVs/HVs is different from those available on the market today and can reduce or eliminate noise which will help make vehicles safe.

3.6. Future Development Trends for Grease Used in EVs/HVs

As EVs are different from ICEs, it is important to know their different future requirements. Of the multitude of greases on the market, lithium greases are a top choice because of their high adherence, non-corrosiveness, and moisture resistance—all of which allow them to be compatible with and protect components [28,29]. Lithium grease has shown to impart the advantages of high adherence, non-corrosiveness, and moisture resistance, making them compatible with several OEM applications such as EVs/HVs. Aluminum and urea greases perform well too; however, their production is associated with hazardous processing and constraints on process balance.

It is predicted that the use of eco-friendly and bio-degradable greases will increase. Additionally, corrosion protection, low-temperature performance, and water resistance, among others, will rise in interest [29–31]. It is also important to consider that it is more favorable to produce a grease with low torque functionality through thickeners, base oils, and additives while ensuring that the properties of the electrical and surfaces remain unaltered. In the future, it is likely that instead of general greases for all types of EVs, custom-made ones will be favored due to the variabilities in EV designs and the factors they will bring into the formulation of greases [18,32]. While PU greases are currently not commonly used in EVs, they might be of interest in the future due to their lifelong sealing functions, long-life properties at high temperatures, and low noise characteristics. PU grease formulations have inherent oxidation inhibition chemistry as part of the thickener, leading to long life at elevated temperatures. When coupled with PAO stocks, these products can contribute to significant improvements in grease life at elevated bearing temperatures. Furthermore, improved filtration practices and thickener reaction control have contributed to lower noise properties by controlling particles' size and overall particle contamination. This leads to quieter noise levels in rotating bearings.

Desired lubricant properties in EVs/HVs [15,33,34] have been grouped into a table as shown in Table 2. All the required lubricant properties that need to test their performance characteristics are compared with the conventional lubricants used in the ICEs. This table indicates that these newly developed tests such as electrical conductivity, thermal conductivity, extreme speed, and bearing protection capability are crucial performance characteristics to meet HV or EV performance requirements.

Table 2. Desired lubricant properties in electric vehicles (EVs)/hybrid vehicles (HVs) compared with in internal combustion engine vehicles (ICEVs) (adapted from [15,33,34]).

Serial Number	Lubrication Parameter [1]	ICEV Requirement	EV Requirement	Location [2]
1	Acid value	Should be within acceptable limits to avoid corrosion (ASTM D 974 and DIN 51558 may be referred)	Should be extremely low compared with ICE to avoid any corrodibility of polymer parts or motor components	All
2	Anti-foaming	Should have anti-foaming properties	Anti-foaming is highly desirable at high entertainment speeds of lubricant due to higher susceptibility to foaming	2–5
3	Corrosion resistance	Should not corrode the metallic parts	Should be highly compatible with polymers and metal working parts and not lead to corrosion	All
4	Degradability	Resistance to thermal degradation	Resistance to thermal and electrical degradation	2
5	Density	Moderate- to high-density oils preferred	Low-density lighter oil preferred	3–5
6	Dielectric strength	Moderate to low is acceptable	Should not undergo dielectric breakdown under a high electric field	2
7	Electrical conductivity	Should have a good insulating property	Should be moderately conductive to remove static charges but not highly conductive which can cause short-circuiting	2
8	Flammability	Should not be flammable under high heat	Should not be inflammable under high heat and electrical discharge conditions	All
9	Flash point	High flash and fire points are desired	The flash and fire points need to be very high compared with ICE	All
10	Heat transfer	Should have moderate to high heat transfer coefficient to dissipate engine heat	Should have a high transfer coefficient and cooling property to remove large heat generated due to high motor speed	1,6
11	Longevity	Should last an acceptable life time, needs refilling and oil change. Many new models are designed for fill-for-life	Long life or fill-for-life preferred	All
12	Pour point	Low to moderate pour point of lubricant is acceptable depending on geography	Pour point for EV lubricant, for the same geographical location, would be the same as that of an ICEV lubricant. However, low pour point is desired for operability at wider environmental conditions at the global scale for new EV designs	All
13	Temperature stability	Should be stable in the working temperature range of the engine	Should be stable under a wide temperature range and be able to withstand sudden and multiple thermal shocks and temperature gradients	2–5
14	Viscosity	High viscosity preferred to support the bearing load	Low viscosity preferred for better cooling performance (Van Rensselar, 2019)	1–6
15	Volatility	Should not be volatile under the influence of thermal and pressure variations of the engine	Should have even better volatility resistance than ICE oils considering frequent start stops and shock loads	All
16	Water resistance	Should have water resistance and a hindrance to water in oil type emulsion formation	Should have higher water resistance and hydrophobicity to avoid electro-wetting	All
17	Wear resistance	Should have anti-wear properties	Should not lead to wear of components at high temperature and electric field conditions	2–5

[1] In reference to specific lubrication types required for vehicles in Figure 3B. [2] The location numbers are in reference to the labels present in Figure 3B.

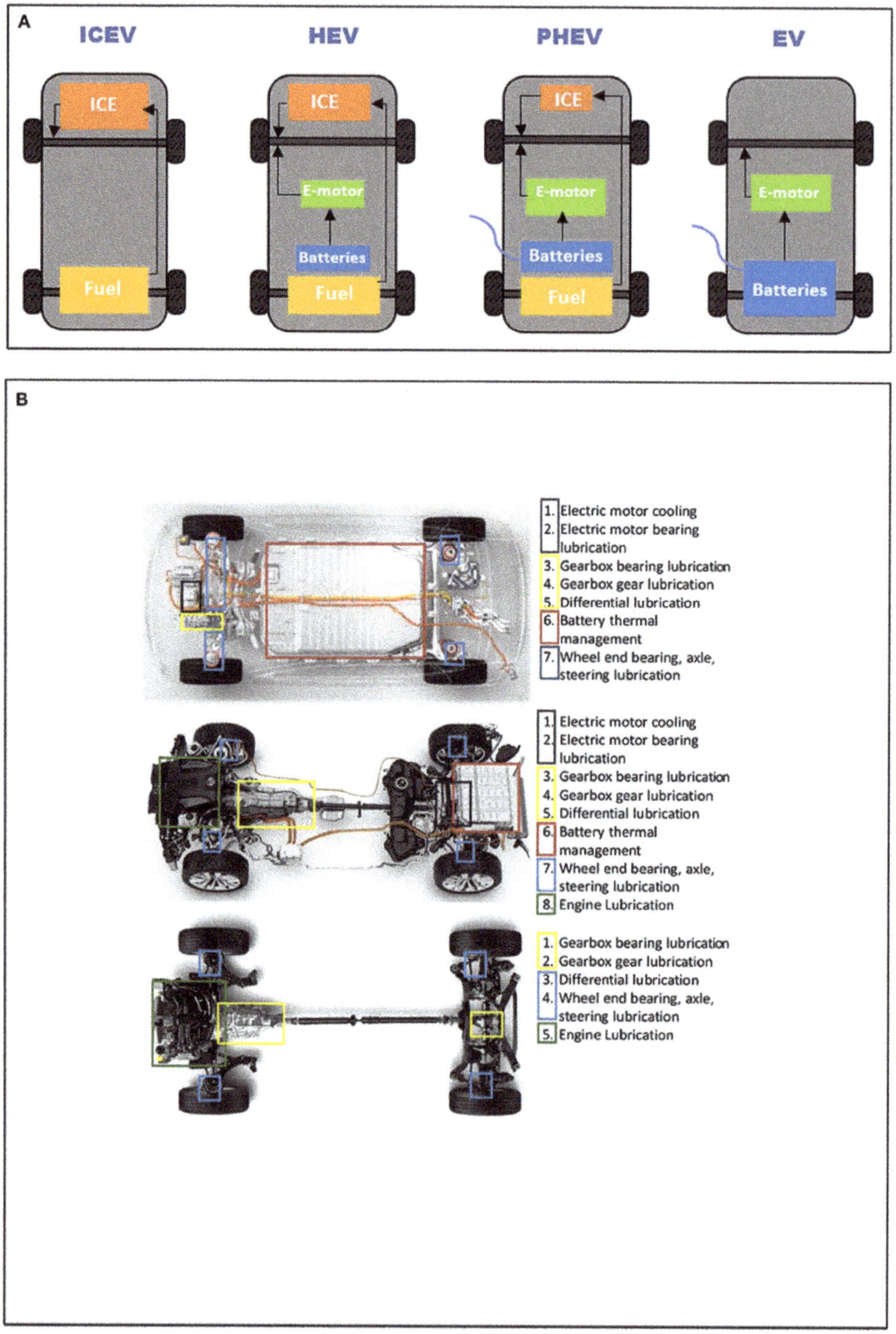

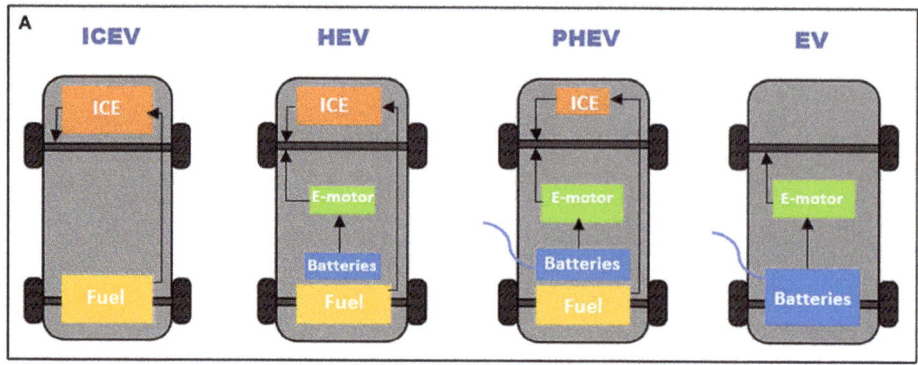

Figure 3. (**A**) Comparison of battery and ICE sizes and (**B**) an image of the main areas where lubrication is required in an EV (Tesla), hybrid electric vehicle (HEV) (Volkswagen NetCarShow), and ICEV (Subaru Forester showroom), respectively [15].

3.7. ASTM Standard Test Development for Grease Used in EVs or HVs

Various ASTM standard tests [34] have been developed to ensure the required performance specifications for grease operation in EVs or HVs. Among these ASTM standard tests, the most important test parameters are shown in Table 3. For example, oil viscosity is developed according to the load, speed, and operating temperature of the application [32]. While viscosity should be reduced to minimize friction loss, too low of a viscosity hinders durability and causes the lubricant to leak out of the bearings. This brings into play oxidation properties and dropping point when in extreme temperatures. Oxidation, enhanced by spark discharges, deteriorates the oil and increases the chances of sludge buildup, which hinders thermal control from the motor. Additives are added to modify these properties, but some may be counterintuitive and shorten the life span of the grease. Lastly, the lubricant must maintain electrical properties such as volume resistivity, dissipation factor, and dielectric strength to avoid electrical losses in the system [32]. Overall, the lubricant must be formulated to balance all these requirements. Table 4 depicts specific grease properties that must be tested today for operation in EVs or HVs.+

Table 3. Laboratory test specifications for electrical properties (adapted from [34]).

Test Specification	Required Characteristics
ASTM D149	Dielectric strength
ASTM D257	DC resistance or conductance
ASTM D1816	Dielectric breakdown voltage of insulating liquids
ASTM D2624	Electrical conductivity of aviation and distillate fuels
ASTM D4308	Electrical conductivity of liquid hydrocarbons
GRW or SKF Be Quiet	Grease noise bearing test (company tests)
To be developed	Loss tangent tan δ
To be developed	Relative permittivity

Table 4. ASTM laboratory test specifications for functional driveline fluids or greases (adapted from [34]).

Required Characteristics	Test Specification
DIN 51819 FAG FE8 (wear of rollers)	Superior wear properties under accelerated rolling contact fatigue
DIN 51821 FAG FE9	Grease life/oxidation stability
ASTM 02265 Dropping Point	High operating temperatures
IP 121, ASTM D1742, ASTM D6184	Excellent oil release properties
ASTM D4170, D7594 Fretting	Fretting wear resistance and low friction
SNR FEB2 (company test)	False brinelling test
ASTM 06138 Anti-Rust Test	Corrosion resistance
ASTM D1478 Col start torque	Low-temperature torque
ASTM D4950 NLGI Grade	Consistency
ASTM D217 Worked Cone Penetration (100Kx)	Mechanical/work stability
ASTM D1831 Grease roll stability	Resistance to physical degradation
ASTM D2266, D2596, D7421, D5706	Extreme pressure (EP) properties
ASTM D1264 Water wash out	Water resistance
ASTM D4289 Elastomer compatibility	Seal compatibility
Courtesy of Kuldeep Mistry and The Timken Co.	

In addition, ASTM Corrosion Tests [32] for grease have been developed such as the ASTM D4048 Copper Corrosion from Grease, D5969 Corrosion of Grease in Sea Water, and D1743—Corrosion Preventive Properties of Lubricating Greases. ASTM offers other basic lab test methods for specific properties related to electric fields and thermal cooling conditions, which have yet to be reviewed and assessed for greases used in EVs or HVs. A future publication for further standard test methods will be addressed.

4. Conclusions

Greases have been used in manufacturing and automotive industries and have been rapidly evolving in recent decades to meet the demands of modern automobiles. While some greases have specific niche applications, the most common ones are those that are versatile and can be used in a wide variety of situations. Lithium-based greases, for example, are some of the most popular greases on the market due to their wide temperature tolerance and high dropping point. The reason for their popularity is due to the effective solution for a major cause of grease failure—high temperature. Greases operating at high temperatures and loads for ICEVs' components cannot be applied for EVs. EVs or HVs, while they also produce heat, prefer to operate in cooler conditions, requiring specific greases that can function at those temperatures as well.

Furthermore, one important aspect of EVs or HVs not present in ICEVs is the specific performance requirements of electrification components. These create electromagnetic fields and electrical currents, leading to potential issues as the field can transfer energy without the need for a physical medium. It became common for greases—especially thermally and electrically non-conductive ones—to absorb too much of the energy and eventually degrade due to oxidation and evaporation. The difference in operating conditions leads to a difference in requirement of lubricants in the two vehicles. The preferred cooler temperature of EVs leads to the pursuit of both a lower viscosity and density of lubricants while the opposite is true for ICEVs. While ICEVs are abundant today, EVs will undoubtedly become the future and their lubrication and grease components will be a focal point.

The major differences between EVs and conventional ICEVs can be grouped into the following technical areas: energy efficiency, noise, vibration, and harshness (NVH) issues and the presence of electrical current and electromagnetic fields from electric modules, sensors and circuits, and bearing lubrication. Additional considerations include thermal transfer, seals, corrosion protection, and materials' compatibility. The authors will review the future development trends of EVs/HVs on driveline lubrication and thermal management requirements. The future development of electric vehicles will globally influence the selection and development of gear oils, coolants, and greases, as they will be in contact with electric modules, sensors, and circuits and will be affected by electrical current and electromagnetic fields.

The increasing presence of electrical parts in EVs/HVs will demand the corrosion protection of bearings and other remaining mechanical components. Thus, it is imperative that specialized greases should be explored for specific applications in EVs/HVs to ensure maximum protection from friction, wear, and corrosion to guarantee the longevity of the operating automobile.

Author Contributions: Conceptualization, R.S., R.C., S.T. and R.M.; literature review and formal analysis, R.C. and S.T.; writing—original draft preparation, S.T. and R.C.; writing—review and editing, R.S., S.T. and R.M.; revisions and supervision, R.S., S.T. and R.M. All authors have read and agreed to the published version of the manuscript.

Funding: This research received no external funding.

Institutional Review Board Statement: Not applicable.

Informed Consent Statement: Not applicable.

Data Availability Statement: Data from reference [14] (Figures 1 and 2) and reference [15] (Table 2 and Figure 3) are publicly available from https://www.semanticscholar.org/paper/Decomposition-Kinetic-of-Greases-by-Thermal-Rhee/80d4d8f1ac0213173967ad1b6b4d15c3689f9026 (accessed on 2 March 2021) and https://doi.org/10.3389/fmech.2020.571464 (accessed on 2 March 2021), respectively. Data from Table 1 (reference [8]) and Tables 3 and 4 (reference [34]) have restricted access. Data from [8] is available at https://www.elsevier.com/books/the-tribology-handbook/neale/978-0-7506-1198-5 (accessed on 2 March 2021) and [34] is available at https://www.astm.org/DIGITAL_LIBRARY/MNL/PAGES/MNL6220121208809.htm (accessed on 2 March 2021).

Conflicts of Interest: The authors declare no conflict of interest.

References

1. Ahmed, N.; Nassar, A. Lubrication and Lubricants. Available online: https://www.intechopen.com/books/tribology-fundamentals-and-advancements/lubrication-and-lubricants (accessed on 9 December 2020).
2. Greases for Your Vehicle. Available online: https://thelubepage.com/amsoil-magazine-articles/magazine-articles/articles/grease-for-your-vehicle (accessed on 10 December 2020).
3. Wright, J. Grease Basics. Available online: https://www.machinerylubrication.com/Read/1352/grease-basics (accessed on 10 December 2020).
4. Shah, R.; Wong, H.; Law, A.; Woydt, M. The New Age of Lubricants for Electric Vehicles. Available online: https://www.electrichybridvehicletechnology.com/features/the-new-age-of-lubricants-for-electric-vehicles.html (accessed on 3 December 2020).
5. Farfan-Cabrera, L. Tribology of Electric Vehicles: A Review of Critical Components, Current State and Future Improvement Trends. Available online: https://www.sciencedirect.com/science/article/pii/S0301679X19303433?via%3Dihub (accessed on 10 December 2020).
6. Choosing the Right Grease Thickening System. (n.d.). Available online: https://www.nyelubricants.com/choosing-the-right-grease-thickening-system (accessed on 10 December 2020).
7. Klamann, D. *Schmierstoff und Verwandte Produkte*; VCH-Verlag: Weinheim, Germany, 1982.
8. Neale, M.J. C4—Greases. In *The Tribology Handbook*, 2nd ed.; Butterworth-Heinemann: Oxford, MA, USA, 1995; pp. C4.1–C4.4.
9. Tung, S.C.; Woydt, M.; Shah, R. Global Insights on Future Trends of Hybrid/EV Driveline Lubrication and Thermal Management. *J. Front. Mech. Eng. Tribol.* **2020**, *6*, 74. [CrossRef]
10. Noria Corporation. Advantages of Using Polyurea Grease. Available online: https://www.machinerylubrication.com/Read/31367/using-polyurea-grease (accessed on 15 December 2020).
11. Monkey, G. Complete Grease Guide—Workshop Greases. Available online: https://www.greasemonkeydirect.com/blogs/news/complete-grease-guide-workshop-greases (accessed on 15 December 2020).
12. Dowson, D. *'Elastohydrodynamic Lubrication,' Interdisciplinary Approach to the Lubrication of Concentrated Contacts*; NASA Spec. Pub-237; National Aeronautics and Space Administration: Washington, DC, USA, 1970; p. 34.
13. Hamrock, B.J.; Dowson, D. Isothermal Elastohydrodynamic Lubrication of Point Contacts, Part III, Fully Flooded Results. *ASME J. Lubr. Technol.* **1997**, *99*, 264–276. [CrossRef]
14. Rhee, I. Decomposition Kinetic of Greases by Thermal Analysis. In Proceedings of the 74th NLGI Annual Meeting, Scottsdale, AZ, USA, 27 April 2007.
15. Chen, Y.; Jha, S.; Raut, A.; Zhang, W.; Liang, H. Performance Characteristics of Lubricants in Electrical and Hybrid Vehicles: A Review of Current and Future Needs. *J. Front. Mech. Eng. Tribol.* **2020**, *6*, 82.
16. He, F.; Xie, G.; Luo, J. Electrical bearing failures in electric vehicles. *Friction* **2020**, *8*, 4–28. [CrossRef]
17. Chau, K.T.; Chan, C.C. Emerging Energy-Efficient Technologies for Hybrid Electric Vehicles. *Proc. IEEE* **2007**, *95*, 821–835. [CrossRef]
18. Pettersson, A. High-performance base fluids for environmentally adapted lubricants. *Tribol. Int.* **2007**, *40*, 638–645. [CrossRef]
19. Callen, H.B. *Thermodynamics and an Introduction to Thermostatics*; Wiley: New York, NY, USA, 1985.
20. Gedde, U. *Polymer Physics*, 1st ed.; Springer Science & Business Media: Dordrecht, The Netherlands, 1995.
21. Jin, H.; Andritsch, T.; Tsekmes, I.A.; Kochetov, R.; Morshuis, P.H.; Smit, J.J. Properties of mineral oil based silica nanofluids. *IEEE Trans. Dielectr. Electr. Insul.* **2014**, *21*, 1100–1108.
22. Shaikh, S.; Lafdi, K.; Ponnappan, R. Thermal conductivity improvement in carbon nanoparticle doped PAO oil: An experimental study. *J. Appl. Phys.* **2007**, *101*, 064302. [CrossRef]
23. Barbés, B.; Páramo, R.; Blanco, E.; Pastoriza-Gallego, M.J.; Pineiro, M.M.; Legido, J.L.; Casanova, C. Thermal conductivity and specific heat capacity measurements of Al_2O_3 nanofluids. *J. Therm. Anal. Calorim.* **2013**, *111*, 1615–1625. [CrossRef]
24. Chen, Y.; Renner, P.; Liang, H. Dispersion of nanoparticles in lubricating oil: A critical review. *Lubricants* **2019**, *7*, 7. [CrossRef]
25. Dai, W.; Kheireddin, B.; Gao, H.; Liang, H. Roles of nanoparticles in oil lubrication. *Tribol. Int.* **2016**, *102*, 88–98. [CrossRef]
26. Håkansson, B.; Andersson, P.; Bäckström, G. Improved hot-wire procedure for thermophysical measurements under pressure. *Rev. Sci. Instrum.* **1988**, *59*, 2269–2275. [CrossRef]
27. Nagasaka, Y.; Nagashima, A. Simultaneous measurement of the thermal conductivity and the thermal diffusivity of liquids by the transient hot-wire method. *Rev. Sci. Instrum.* **1981**, *52*, 229–232. [CrossRef]
28. Andrew, J.M. The future of lubricating greases in the electric vehicle era. *Tribol. Lubr. Technol.* **2019**, *75*, 38–44.
29. Tung, S.C. The Impact of Advanced Propulsion System on Thermal Management and Lubrication Requirements. In Proceedings of the China International Lubricants Conference and Technology Exhibition (INTEX) Keynote Speaker, Guangzhou, China, 21–23 August 2019.
30. Tung, S.C. Overview of Energy Innovation Technology and Future Development Trends Responding to Improving Energy Efficiency Requirements and Green Environmental Opportunities. Invited Speaker. In Proceedings of the SAE China International Congress, Beijing, China, 25 October 2019.
31. Brodie, G. Energy Transfer from Electromagnetic Fields to Materials. Available online: https://www.intechopen.com/books/electromagnetic-fields-and-waves/energy-transfer-from-electromagnetic-fields-to-materials (accessed on 15 December 2020).
32. Suzumura, J. Prevention of electrical pitting on rolling bearings by electrically conductive grease. *Q. Rep. RTRI* **2016**, *57*, 42–47. [CrossRef]

33. Peskoe-Yang, L. Electric vehicles make grease's future uncertain. *Tribol. Lubr. Technol.* **2020**, *76*, 24–25.
34. Tung, S.C.; Totten, G. Chapter 9—Testing of Evaluation of Lubricating Grease for Rolling Element Bearings of Automotive System. In *Automotive Lubricants and Testing*; ASTM International: West Conshohocken, PA, USA, 2012; pp. 137–156.

Case Report

On the Fictitious Grease Lubrication Performance in a Four-Ball Tester

Sravan K. Joysula, Anshuman Dube, Debdutt Patro and Deepak Halenahally Veeregowda *

Application Development Laboratory, Ducom Instruments, 9747 AA Groningen, The Netherlands; sravan.k@ducom.com (S.K.J.); anshuman@ducom.com (A.D.); debdutt.p@ducom.com (D.P.)
* Correspondence: deepak.v@ducom.com

Abstract: The extreme pressure (EP) behavior of grease is related to its additives that can prevent seizure. However, in this study following ASTM D2596 four-ball test method, the EP behavior of greases was modified without any changes to its additive package. A four-ball tester with position encoders and variable frequency drive system was used to control the speed ramp up time or delay in motor speed to demonstrate higher grease weld load and lower grease friction that were fictitious. A tenth of a second delay in speed ramp up time had showed an increase in the weld load from 7848 N to 9810 N for grease X and 6082 N to 9810 N for grease Y. Further increase in the speed ramp up time to 0.95 s showed that the greases passed the maximum load of 9810 N that was possible in the four-ball tester without seizure. The mechanism can be related to the delay in rise of local temperature to reach the melting point of steel required for full seizure or welding, that was theoretically attributed to an increase in heat loss as the speed ramp-up time was increased. Furthermore, the speed ramp up time increased the corrected load for grease X and Y. This resulted in lower friction for grease X and Y. This fictitious low friction can be attributed to decrease in surface roughness at higher extreme pressure or higher corrected load. This study suggests that speed ramp up time is a critical factor that should be further investigated by ASTM and grease manufacturers, to prevent the use of grease with fictitious EP behavior.

Keywords: four-ball tester; heat dissipation; speed ramp up; ASTM D2596; grease weld load; grease friction

Citation: Joysula, S.K.; Dube, A.; Patro, D.; Veeregowda, D.H. On the Fictitious Grease Lubrication Performance in a Four-Ball Tester. *Lubricants* **2021**, *9*, 33. https://doi.org/10.3390/lubricants9030033

Received: 15 January 2021
Accepted: 9 March 2021
Published: 12 March 2021

Publisher's Note: MDPI stays neutral with regard to jurisdictional claims in published maps and institutional affiliations.

Copyright: © 2021 by the authors. Licensee MDPI, Basel, Switzerland. This article is an open access article distributed under the terms and conditions of the Creative Commons Attribution (CC BY) license (https://creativecommons.org/licenses/by/4.0/).

1. Introduction

Several standards have been developed to establish grease performance and load carrying capability [1–4]. Both good [5] and poor correlation [6] has been reported within the different methods using similar lubricants. These have been explained by either similarities or differences between the contact geometries, configurations, and scuffing detection criteria. One such method, the four-ball tester [1,7] having high precision, has been widely used to manage lubricants' batch production quality control. It has helped scientists to select additives, both conventional as well an environment friendly nanoparticle, for extreme pressure, wear prevention and antifriction grease behavior [5,8,9]. Often, short duration four ball tests of 10 s or 60 s is found to be effective in determining the competing effect of additive molecules in surface deposition, tribofilm formation and protection against friction, seizure, and wear. These short duration test methods are standardized and frequently validated by D02 committee in ASTM—American Standards for Testing Materials. ASTM D2596-15 and ASTM D2783-15 are such test methods that are widely practiced by lubricant manufacturers to determine the extreme pressure (EP) behavior of greases. These standards offer vital information about seizure prevention by EP additives at a given load that is known as "weld load" [1]. Almost every grease specification sheet carries the four ball weld load data, as it is intended to help the consumers to choose the best grease to prevent seizure of critical components under starved lubrication conditions. Therefore, weld load data is important for both the grease consumers and manufacturers, with a higher weld load indicating better capability against failure. However, this critical data could be "manipulated" within the scope of ASTM D2596-15 or D2783-15, and it can "trick" the consumers to use the lubricant that can be detrimental to critical components.

In ASTM standards like D2596 or D2783 the lubricant is compressed and sheared between the four balls (top ball and three bottom balls) for 10 s (see Figure 1). The mean speed of the top ball is fixed at 1770 rpm. This test is repeated at every load stage from 6 kg to 800 kg or until the full seizure.

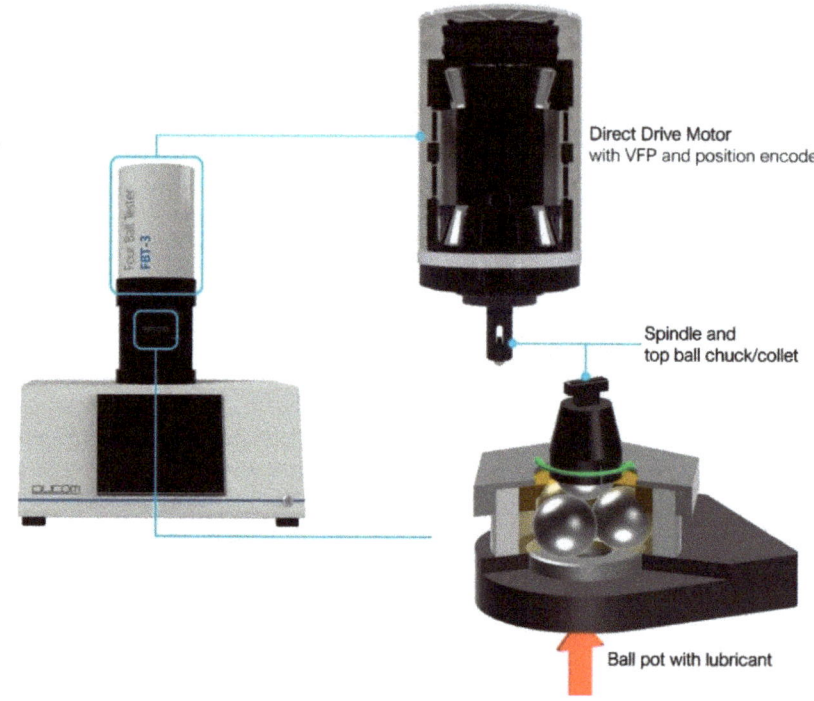

Figure 1. Four ball tester and four ball test configurations in the ball pot.

Seizure is represented by the sudden jump in the motor torque due to melting and fusion of steel balls followed by welding of four balls. To achieve higher weld load the lubricants are formulated with high performing EP additives [5]. In contrary, the desired weld load could also be achieved by tuning the speed "ramp up time" in the four-ball tester. Speed ramp up time is the delay in time taken to reach the mean speed of 1770 rpm. Although the mean speed is described in the ASTM standards, the ramp up time is not mentioned, that could result in fictitious grease lubrication performance. There have been four ball test reports that showed effect of speed on grease wear [10] and effect of delay in applied load on lubricant scuffing loads [11]. However, there are no reports on the effect of ramp up time on grease seizure load or weld load.

In this study, we have developed a four-ball test method to control and measure the speed ramp up time or delay in motor speed in the four-ball tester, whose effect on weld load, friction and wear is investigated for two types of greases. And we propose mechanisms that can explain the changes in weld load and friction due to delay in reaching the mean speed.

2. Materials and Methods

2.1. Greases

Two commercial high weld load greases that can offer protection against wear, scuffing and pitting in gear drives were used in this study. These greases are labelled as grease X and grease Y. Both the greases were NLGI grade 00, they had same density of 0.92 g/cc at 20 °C, kinematic viscosity of 500 cSt at 100 °C, flash point greater than 200 °C, thermal stability of tribofilms was equal to 120 °C and FZG scuffing load stage was equal to or better than 12. There was no information about composition of extreme pressure additives used in these greases. And it is not critical for this study because we are focused on consequences of the test method on grease lubrication behavior. Moreover, we are not investigating lubrication mechanism based on the composition of greases.

2.2. Controlling the Speed Ramp Up Time and Load in a Four-Ball Tester

Computer controlled and automated four ball tester (Model–FBT3) from Ducom Instruments^{TM} (Groningen, The Netherlands) was used in this study (see Figure 1). A variable speed direct drive motor without any belt or pulley arrangements was used to control the speed between 300 rpm to 3000 rpm. Speed ramp up time, that is time delay in motor speed to reach 1770 rpm starting from 0 rpm was controlled using the position encoders. Position encoders can precisely identify the angular position of the spindle in the motor. And they were in closed loop with the variable frequency drive system that controlled the flow of current to the motor, and the motor speed. Variable frequency drive ensures that the speed ramp up time is not affected by starting motor torque that is crucial for ASTM D2596. The direct drive motor without any gear box was compatible with peak load of 10,000 N, to sustain maximum torque at zero speed. Safety controls were used to prevent the overflow of current to the motor at the peak torque operating conditions. The labview based WinDucom software was used to set the desired speed ramp up time for each test. In this study we chose speed ramp up time 0.15 s, 0.25 s and 0.95 s, that is the time delay for motor to reach a preset mean speed of 1770 rpm (see Figure 2A). The above time intervals were chosen considering the motor capabilities and technology used in commercial four ball testers.

The data acquisition and display system in WinDucom software allowed the user to view and store the real time changes in speed profiles.

Ducom four ball tester is equipped with an automated pneumatic loading system that can control the actual load between 100 N to 10,000 N. The standard error at 10,000 N was ±20 N or 0.2%. The test balls were preloaded to a desired load at zero rpm and the load was maintained stable during the spindle rotation for the entire test duration of 10 s and at all the different speed ramp up time (see Figure 2B). The data acquisition and display system in WinDucom software allowed the user to view and store the real time changes in load profiles.

2.3. Pass Load and Weld Load

According to ASTM D2596, the grease is packed into the ball pot with three stationary steel balls (supplied by SKF, E-52100, with diameter of 12.7 mm, Grade 25 extra polish, hardness 65 to 66 HRC) at a temperature of 27 ± 8 °C, the top steel ball connected to the motor is brought in contact with the bottom three steel balls at a fixed load. The top steel ball rotates at a mean speed of 1770 ± 60 rpm for a test duration of 10 s. If there was no welding of the test balls, the load is increased to the next load step, using a look up chart for load steps given in the ASTM D2596.

The weld load is the load step at which the test balls local temperature reached the melting point of steel, that fused the four balls. At this point the friction torque sensor in the four-ball tester exceeds the safety value and shuts down the motor. This represents the failure by grease lubricants to prevent seizure. The load step prior to the weld load is the pass load. The pass load represents the state of the grease lubricant after incipient seizure and before the full seizure. The pass load and weld load for grease X and Y was measured at a speed ramp up time of 0.15 s, 0.25 s, and 0.95 s. There were new steel balls used for each test.

The cleaning procedures in this study followed the ASTM D2596.

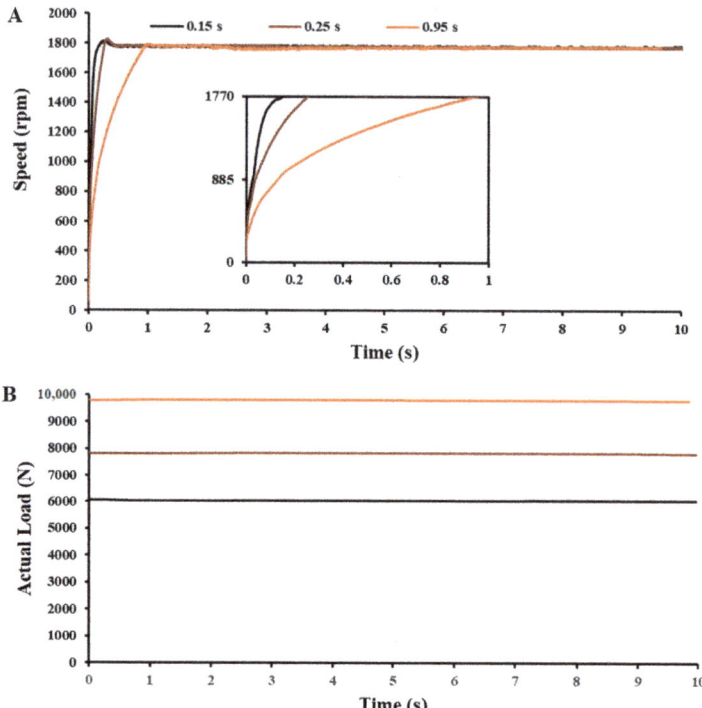

Figure 2. Measurement and controls for speed ramp up time and normal load in a Ducom four-ball tester. Real time changes in the speed profiles (**A**) and load profiles (**B**) at ramp up time of 0.15 s, 0.25 s and 0.95 s. Note: Ramp up time is the time taken to reach the average speed of 1770 rpm (average speed follows ASTM D2596-15, see the figure inset) and the motor was designed for full torque or load at zero speed.

2.4. Ball Mean Wear Scar Diameter and Corrected Load

At every pass load there is severe wear on the three test balls in the ball pot. The mean value of the wear scar diameter on these three test balls can be measured using a microscope to determined ball mean wear scar diameter. The corrected load is a pass load that is compensated with the wear. It is calculated by multiplying the pass load with the ratio of Hertzian contact diameter to ball mean wear scar diameter. The corrected load was determined for grease X and Y at a speed ramp up time of 0.15 s, 0.25 s, and 0.95 s.

2.5. Friction Coefficient

The friction measuring system in the four-ball tester has been extensively described in the US patent US 2017/0176319 A1. Friction torque was measured using a load cell, that was in contact with the moment arm fixed to the ball pot in the four-ball tester. Coefficient of friction was calculated from the measured friction torque and applied load as per ASTM D5183 [12] The data acquisition and display system in WinDucom software allowed the user to view and store the real time changes in friction coefficient profiles. The average friction coefficient was calculated by determining the mean of all the friction coefficient values acquired during a pass load test for grease X or grease Y.

3. Results

3.1. Changes in Pass Load, Corrected Load, and Weld Load

Speed ramp up time had an influence on the pass load, corrected load and weld load of grease X and Y (see Figure 3). As shown in Figure 3A, the pass load and corrected load increased with an increase in ramp up time, they were in power-law relationship for grease X. A similar trend was observed for grease Y however the power law relationship was weaker compared to grease X (see Figure 3B). As shown in Figure 3C, the weld load for grease X was 7848 N, and the grease Y had a lower weld load of 6082 N, at the ramp up time of 0.15 s. At higher ramp up time of 0.25 s, that is a tenth of second delay in motor speed, the grease X and Y had the same weld load of 9810 N. Grease X and Y had passed maximum load of 9810 N in the four-ball tester at a speed ramp up time of 0.95 s.

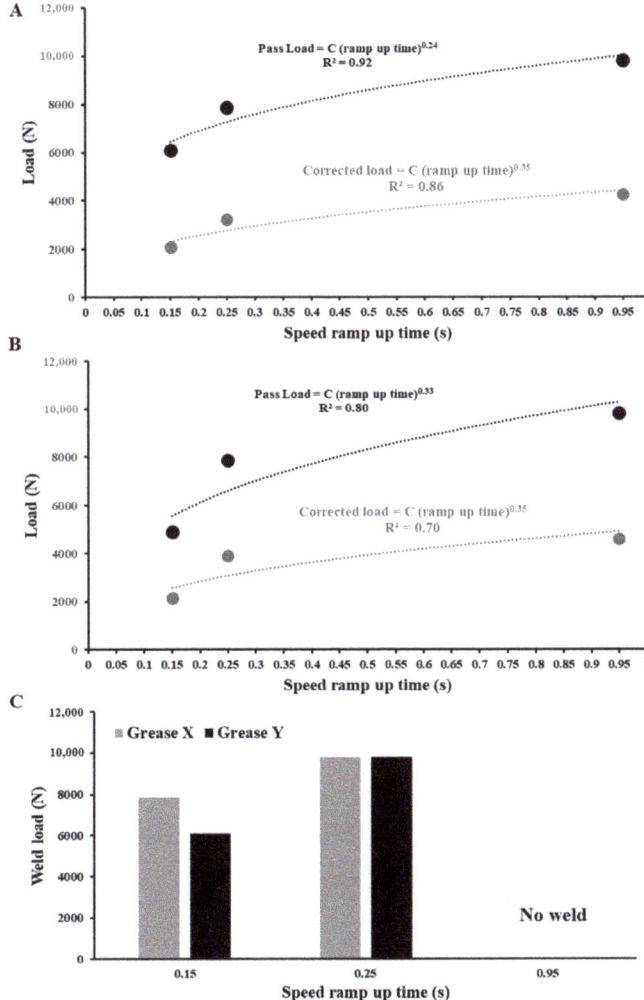

Figure 3. Influence of speed ramp up time on corrected load, pass load and weld load for grease X and Y. A correlation between speed ramp up time and load (corrected load and pass load) for grease X (**A**) and grease Y (**B**), and the changes in weld load due to ramp up time for grease X and Y (**C**). All the measurements were in duplicate.

3.2. Changes in Friction Coefficient

Friction coefficient for the greases was in the range of 0.04 to 0.07, and it was affected by the changes in corrected load driven by changes in the speed ramp up time (see Figure 4). As shown in Figure 4A, the friction coefficient increased and then decreased before reaching a stable plateau. The average friction coefficient for grease X and Y decreased as there was an increase in corrected load, that was driven by an increase in speed ramp up time (see Figure 4B).

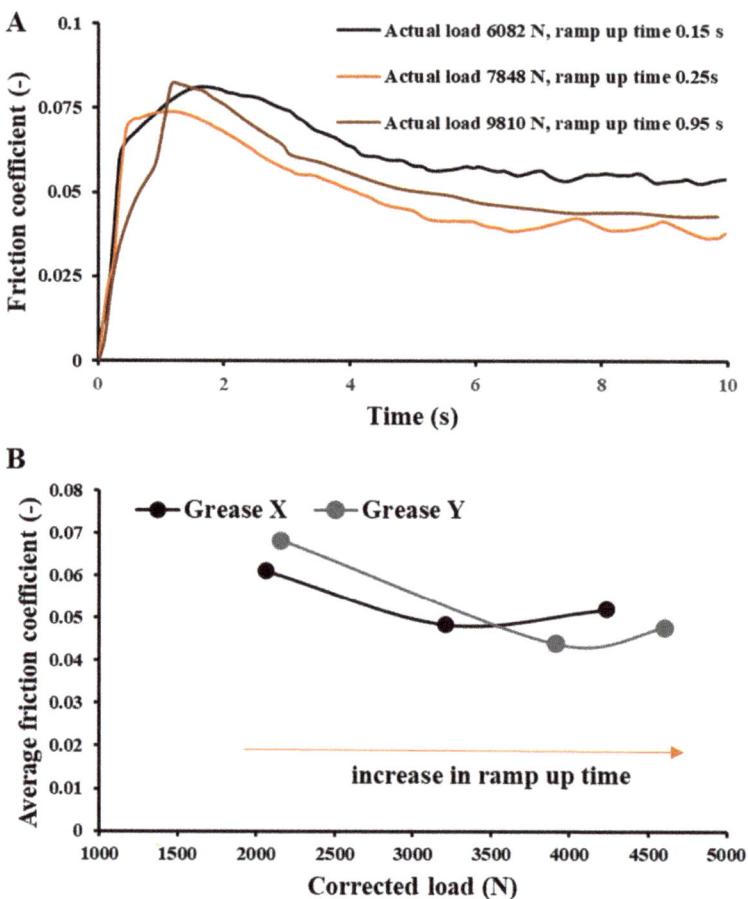

Figure 4. Influence of speed ramp up time on friction for grease X and Y. Real time changes in the friction profiles for grease X (**A**) and the relationship between average friction coefficient and corrected load due to an increase in ramp up time (**B**). All the measurements were in duplicate.

4. Discussion

This is the first study that has experimentally shown that the grease performance as determined by weld load and friction coefficient can be influenced by the speed ramp up time or delay in motor speed—an unknown four-ball test parameter until now. It is fascinating to see that a tenth of second delay in motor attaining the mean speed of 1770 rpm, that is an increase in speed ramp up time from 0.15 s to 0.25 s, had increased the weld load and decreased the friction coefficient for both the greases. Grease weld load represents the inability of EP additives in grease to form a stress activated antiseizure tribofilms that resist sudden rise of local temperature on the steel balls. Seizure occurs when the local flash temperature, reaches the melting point of steel (approximately 1400 °C) resulting in welding of four steel balls [13,14]. A tenth of a second delay in attaining the mean speed means the local temperature is still below the melting point temperature of steel. Therefore, we observed an increase in weld load at higher speed ramp up time. It was possible to never reach the local melting point temperature of steel at a speed ramp up time of 0.95 s. This can be attributed to the timescale over which heat is built-up and dissipated that becomes more important than overall energy available at the contact, referred to as friction power intensity [15]. The frictional power loss (units of $J/s\cdot mm^2$) and heat dissipation rate (units of J/mm^2) differ with the dissipation rate accounting for the timescale the contact is subject to the available frictional power. The frictional power loss and dissipation rate were calculated using the available data in the four-ball tester:

(1) Applied load is equal to 8000 N. The actual normal contact force between all the balls is 9798 N (refer to the Appendices A.1 and A.2)
(2) Spindle rotation is equal to 1770 rpm. This translates to actual sliding speed of 0.678 m/s. (refer to the Appendices A.1 and A.2)
(3) An average friction coefficient value of 0.05 was used based on the data presented in Figure 3A,B and the Appendix A.5 for calculation of coefficient of friction in a four-ball tester.
(4) Initial Hertzian contact diameter, per contact is 0.826 mm. (Normal load of 3266 N per ball, Poisson ratio of 0.27 and elastic modulus of 205 GPa were used). Hence, the total initial contact area for all three balls is 1.61 mm^2.

Frictional power loss per contact area (for all three balls) is [coefficient of friction × actual normal force × sliding velocity]/(total contact area), that results in 206 $J/s\cdot mm^2$.

In Figure 2A (inset) the ramp up time to reach the maximum speed was different, which implies that the available frictional power of 206 $J/s\cdot mm^2$, was dissipated at different rates for the three different speed ramp-up time.

Dissipated heat or heat flow at different ramp-up time can be calculated using the linear method (frictional power × ramp-up time) and the integral of the area within the curve depicted in Figure 2. Please refer to Table 1 for the heat dissipation rates. Both the methods indicate that there was a severe increase in loss of heat (5 times) as the speed ramp up time was increased from 0.15 s to 0.95 s. Thus, the available heat was not adequate for the local conditions to reach to the critical flash temperatures at higher ramp up time thereby leading to a higher weld load. Unfortunately, there are no tools to measure the local temperature rise in a four-ball tester that could have helped us to experimentally confirm this hypothesis.

Table 1. Calculated heat dissipation rates measured for the different speed ramp up time.

Ramp Up Time (s)	Frictional Power Loss per Contact Area ($J/s\cdot mm^2$)	Dissipated Heat during Ramp Up [J/mm^2], Linear Ramp Up	Dissipated Heat during Ramp Up [J/mm^2], Integral of Ramp Up Curves in Figure 2
0.15	206	206 × 0.15 = 31	29
0.25	206	206 × 0.25 = 51.5	39.6
0.95	206	206 × 0.95 = 195.7	142.8

Antiseizure tribofilms formed on the steel surface depends on the type of additive composition in the greases. Grease Y that had poor resistance to seizure compared with grease X had demonstrated a better behavior and it was equal to grease X. This was again triggered by a tenth of a second delay in motor speed. It was interesting to observe that performance of antiseizure tribofilms in grease Y was highly exaggerated by increasing the speed ramp up time compared to grease X. This is an evidence that the tribofilms can react differently to the speed ramp up time. Although it is interesting to investigate the physicochemical nature of these tribofilms, the focus of this paper has been limited to demonstrating the changes in lubricants performance.

Lubricity of grease as determined by friction coefficient was also exaggerated by speed ramp up time. A tenth of a second delay in motor speed had showed a tremendous improvement in grease friction. This can be attributed to an increase in corrected load. At higher load the steel surface and tribofilms are subjected to extreme pressure conditions that can decrease their surface roughness, that could have resulted in decrease in the friction [16]. In line with this mechanism the grease Y that had higher corrected load compared to grease X also showed lower friction compared to grease X.

5. Conclusions

ASTM D2596 was developed with an understanding that seizure prevention by EP lubricants is largely affected by factors like actual load, ball diameter, sliding speed, lubricant temperature and friction that were directly accountable for changes in local temperature. Therefore, it makes sense that these parameters were well described in the standard. However, this study shows that

- Antiseizure and friction performance of grease were improved without modifying the grease chemistry by using the most unknown four-ball test parameter called the speed ramp up time
- Furthermore, it will be crucial to mention the speed ramp up time along with the weld load in the grease data sheet.
- It is important to revise the standard to include the speed ramp up time that had a significant influence on seizure prevention by EP lubricants.

Author Contributions: Conceptualization, D.H.V.; methodology, D.H.V., D.P. and S.K.J.; software, A.D.; validation, D.H.V., D.P. and S.K.J.; formal analysis, D.H.V., D.P. and S.K.J.; investigation, D.H.V., D.P. and S.K.J.; resources, A.D.; data curation, D.H.V., D.P. and S.K.J.; writing—original draft preparation, D.H.V.; writing—review and editing, D.P.; visualization, D.H.V.; supervision, D.H.V. and D.P.; project administration, D.H.V.; funding acquisition, A.D. All authors have read and agreed to the published version of the manuscript.

Funding: This research received no external funding.

Institutional Review Board Statement: Not applicable.

Informed Consent Statement: Not applicable.

Acknowledgments: The authors would like to thank Channabasappa from Kluber Lubrication (Mysore, India) for helping us in selection of high weld load greases and Fabio Alemanno from Ducom Instruments for creating the graphical abstract.

Conflicts of Interest: The authors declare no conflict of interest.

Appendix A. Calculation of Actual Load, Sliding Speed and Coefficient of Friction in Four-Ball Tester

Appendix A.1. Calculation of Subtended Angle in Four Ball Tetrahedral Configuration via CAD Method

For calculating the contact radius of the ball interface, we must know the vertical contact angle, (θ) subtended between the top ball and the bottom three balls. Refer to the image for details. The angle is derived via 3d CAD-based assembly measurements.

Details of the steps used to arrive at the angle are illustrated in the images on the right-hand side.

The first image shows a typical 4-ball assembly as it manifests on a four ball testerHere all the four balls are of identical diameters and measure 12.7 mm.

A cut section view is generated exactly with the center of two balls taken simultaneously. One being the top ball which is held in the collet and the other one is one of the balls which are locked in the ball pot.

A sketch is generated using the references created in the assembly model. The half-angle (θ) here is measured at 35.26°, which is the angle of contact with respect to the axis of rotation and the direction in which the normal load (N) is applied.

Figure A1. Illustration of four ball assembly in four-ball tester.

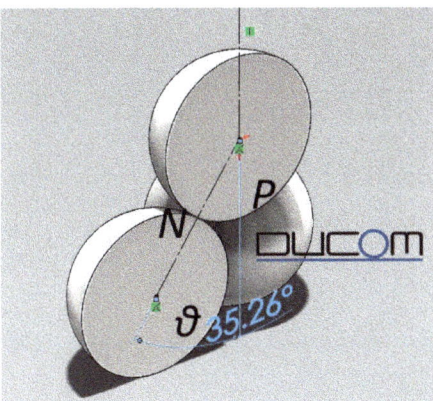

Figure A2. Illustration of top ball in contact with the bottom balls at a fixed angle within a four-ball assembly in four-ball tester.

Appendix A.2. Correlating Applied Test Load, P, with the Local Normal Resultant Force, N, on the Balls

From Section 1, we know that the vertical angle subtended at the contact points of the ball is 35.26° (θ). Knowing this, we can correlate the test load, P, applied to the assembly with the resultant force, N, at the contact. The following steps are used for this:

$$P = \cos(35.26°) \times N$$

$$P = 0.8164N$$

or:

$$N = 1.2247P$$

where, P is the applied load during a test.

Appendix A.3. Measurement of Frictional Torque and Local Frictional Forces, on the Ball

The frictional torque (FT) is measured using a module which consists of a load cell and arm attached to the ball pot and displayed on controller. Using the displayed value, we calculate:

$$FT = f \times r$$

where, f is the frictional force between the contacting balls or:

$$f = FT/r$$

$$f = FT/0.00366$$

$$f = 272.929 \times FT$$

where, r = distance on contact of ball from center (in m):

$$r = \sin 35.26 \times (\text{diameter of the ball}/2)$$

$$r = 0.577 \times (0.0127/2)$$

$$r = 0.577 \times 0.00635 = 0.00366 \text{ m}$$

Appendix A.4. Measurement of Sliding Velocity at the Initial Contact for ASTM D4172 Test

The distance between the central axis of rotation of the top ball and the point of contact between any of the bottom three balls, r, was calculated in Section 3.

Knowing 'r', the sliding velocity at the point of contact can be calculated as:

$$v = \frac{2\pi r N}{60}$$

where v = sliding speed in m/s, r = distance between ball contact and central axis, in m, N = speed of rotation in RPM. For ASTM D2266, N = 1770 RPM. From FBT tetrahedral geometry, d = 0.00366. Hence, sliding speed $v = (2 \times PI \times 0.00366 \times 1770)/60 = 0.678$ m/s.

Appendix A.5. Calculation of Coefficient of Friction Using Applied Load, P, and Measured Friction Torque, FT

Moving on, the coefficient of friction—CoF (μ) is calculated by using the formula:

$$\mu = f/N$$

where, f is the frictional force between the contacting balls, N is the resultant load between the contacting balls.

Now, with the information and derivations from Appendices A.1–A.3, we can calculate CoF:

$$\mu = 272.929 FT/1.2247 P$$

Therefore, CoF at the ball contact can be calculated with the following formula:

$$\mu = 222.854 \times FT/P$$

References

1. ASTM D2596-20, *Standard Test Method for Measurement of Extreme-Pressure Properties of Lubricating Grease (Four-Ball Method)*; ASTM International: West Conshohocken, PA, USA, 2020.
2. ASTM D2783-19, *Standard Test Method for Measurement of Extreme-Pressure Properties of Lubricating Fluids (Four-Ball Method)*; ASTM International: West Conshohocken, PA, USA, 2019.
3. ASTM D2509-20a, *Standard Test Method for Measurement of Load-Carrying Capacity of Lubricating Grease (Timken Method)*; ASTM International: West Conshohocken, PA, USA, 2020.

4. ASTM D5706-16, *Standard Test Method for Determining Extreme Pressure Properties of Lubricating Greases Using a High-Frequency, Linear-Oscillation (SRV) Test Machine*; ASTM International: West Conshohocken, PA, USA, 2016.
5. Fish, G. Extreme Pressure Performance of Greases: Testing and Additive Solutions. *GreaseTech India* **2014**, *4*, 1–12.
6. Hoehn, B.; Michaelis, K.; Doleschel, A. Limitations of Bench Testing for Gear Lubricants. In *Bench Testing of Industrial Fluid Lubrication and Wear Properties Used in Machinery Applications*; ASTM International: West Conshohocken, PA, USA, 2001; pp. 15–32.
7. Boerlage, G.D. Four-ball Testing Apparatus for Extreme Pressure Lubricants. *Engineering* **1933**, *136*, 46–47.
8. Saini, V.; Bijwe, J.; Seth, S.; Ramakumar, S.S.V. Potential exploration of nano-talc particles for enhancing the anti-wear and extreme pressure performance of oil. *Tribol. Int.* **2020**, *151*, 106452. [CrossRef]
9. Kamel, O.; Ali, M. Development and manufacturing an automated lubrication machine test for nano grease. *J. Mater. Res. Technol.* **2020**, *9*, 2054–2062.
10. Bagi, S.D.; Aswath, P.B. Role of MoS2 morphology on wear and friction under spectrum loading conditions. *Lubr. Sci.* **2015**, *27*, 429–449. [CrossRef]
11. Gondal, A.K.; Sethuramiah, A.; Prakash, B. Studies on the tribological behaviour of two oil-soluble molybdenum compounds under reciprocating sliding conditions. *Lubr. Sci.* **1993**, *5*, 337–359. [CrossRef]
12. ASTM D5183-05(2016), *Standard Test Method for Determination of the Coefficient of Friction of Lubricants Using the Four-Ball Wear Test Machine*; ASTM International: West Conshohocken, PA, USA, 2016.
13. Blok, H. The Flash Temperature Concept. *Wear* **1963**, *6*, 483–494. [CrossRef]
14. Wisniewski, M.; Szczerek, M.; Tuszynski, W. The temperatures at scuffing and seizure in a four-ball contact. *Lubr. Sci.* **2004**, *16*, 215–227. [CrossRef]
15. Matveevsky, R.M. The Critical Temperature of Oil with Point and Line Contact Machines. *J. Fluids Eng.* **1965**, *87*, 754–759. [CrossRef]
16. Yamaguchi, K.; Sasaki, C.; Tsuboi, R.; Sasaki, S. Effect of surface roughness on friction behaviour of steel under boundary lubrication. *J. Eng. Tribol.* **1996**, *228*, 1015–1019. [CrossRef]

Technical Note

Quantification of Tackiness of a Grease: The Road to a Method

Emmanuel P. Georgiou [1,*], **Dirk Drees** [1], **Michel De Bilde** [1], **Michael Anderson** [2], **Matthias Carlstedt** [3] **and Olaf Mollenhauer** [3]

[1] Falex Tribology NV, Wingepark 23B, 3110 Rotselaar, Belgium; ddrees@falex.eu (D.D.); mdebilde@falex.eu (M.D.B.)
[2] Falex Corporation, 1020 Airpark Drive, Sugar Grove, IL 60554, USA; manderson@falex.com
[3] Kompass Gmbh, Ehrenbergstraße 11, 98693 Ilmenau, Germany; m.carlstedt@kompass-sensor.com (M.C.); o.mollenhauer@kompass-sensor.com (O.M.)
* Correspondence: egeorgiou@falex.eu; Tel.: +32-164-079-65

Abstract: In this work, we report on the most recent progress in studying temperature influence on tackiness of greases, as well as the reproducibility of the method. Tackiness and adhesion of greases have been identified as key intrinsic properties that can influence their functionality and performance. During the last eight years, a reliable method to quantify the tackiness and adhesion of greases has evolved from an experimental lab-scale set-up towards a standardised approach, including an ASTM method and a dedicated test tool. The performance of lubricating greases—extensively used in diverse industrial applications—is strongly dependent on their adherence to the substrate, cohesion and thread formation or tackiness of the greases. This issue attracts more and more industrial interest as the complexity in grease formulation evolves and it is harder to differentiate between available greases. With this method, grease formulators will have an efficient measurement tool to support their work.

Keywords: grease; tackiness; adhesion

1. Introduction

Tackiness was first identified as an intrinsic characteristic of adhesives materials in the 1940s [1]. Based on this fundamental work, tackiness was defined as the resistance needed to separate two solid surfaces joined by an adhesive layer in its liquid state [1]. In the following years, the term tackiness was adopted in different industrial fields, including that of grease manufacturing. However, depending on the industrial application, it is interpreted in a different way. For example, in the elastomer industry, it is considered as the work required to remove a material from a polymeric film/material [2,3], whereas in the grease industry, tackiness is viewed as the ability of a grease to form threads as it is being pulled apart [3].

Up to now, the most widely used method to evaluate thread formation (tackiness) of greases is the finger test [3]. Based on this empirical method, a grease film is applied between the thumb and index finger, and then they are pulled apart, resulting in the formation of grease threads (Figure 1). According to this test, the longer the threads are, the higher the tackiness of the grease is. The finger test might be a simple-to-perform and zero-cost method, but it is also very empirical and subjective since the results are strongly dependent on the user who is performing the test. Indeed, depending on the amount of grease used and the retraction speed, a different result can be obtained. Another field test that is also used to assess the tackiness of greases is the hammer test [4]. In this method, a 5-pound hammer head is dropped onto a small quantity of grease applied on an anvil. The distance to which the grease is thrown or spreads determines its tackiness (a higher distance means higher tackiness). This method also has the disadvantage that it is not well controlled and that it measures the consistency of the grease upon impact and provides less information about the adhesion or tackiness. In addition, the sensitivity of this approach is rather poor, especially when comparing similar greases. For this reason, there is a strong need to develop a consistent, repeatable method that can quantify tackiness.

Figure 1. Thread formation during physical separation of a grease.

During the last 10 years, more sophisticated experimental methods with different testers have been developed [5–10]. In all cases, the theoretical principle is based on measuring the interaction forces between a probe and a grease layer. From the obtained know-how, the authors have developed a dedicated tester that measures adhesion, separation energy and thread formation (tackiness) of greases from indentation–retraction curves. According to this method, a fixed grease volume is applied on a standardised steel holder. An indenter is gradually brought into contact with the grease until a preset contact load (maximum force in Figure 2). Then, the indenter is gradually moved away from the greased substrate under well-controlled conditions until there is complete physical separation. The pull-off force is identified as the force required to start the physical separation of the indenter from the grease (indicated by the minimum force in Figure 1). As the indenter is retracted, a grease thread forms (Figure 2), which resembles the finger test (Figure 1). The thread length is defined as the displacement from the start of the separation until the force is zero (thread is broken). In addition, the separation energy (Se) is defined as the energy needed to fully separate the indenter from the grease by the pull-off force and is calculated from integration of the area between the start-up of the retraction motion (minimum force) and complete physical separation (Figure 2). Based on this method, tackier greases form longer threads, as with the finger test, whereas the pull-off force relates to the adhesion (also known as stickiness). There is a common misconception that the pull-off force (adhesion) is also an indication of tackiness (thread formation). However, this is not the case, as proved by recent research publications [11,12].

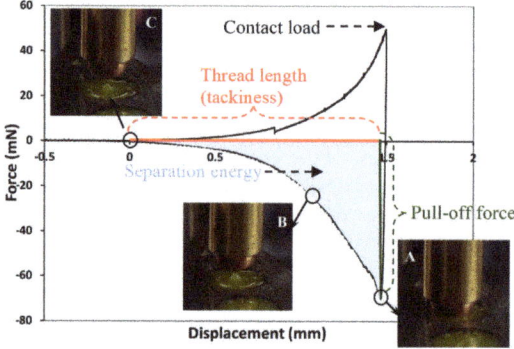

Figure 2. An indicative indentation–retraction curve and measured values.

Despite the significant progress that has been made to quantify grease tackiness, several aspects still need to be addressed. For example, it is known that the performance of greases is temperature dependent. This means that their intrinsic properties, including adhesion, cohesion and tackiness (thread formation), should be also influenced by the temperature in which they operate. Up to now, all tackiness test protocols have been performed at room temperature. Thus, in this work, an attempt is made to investigate the effect of temperature on the adhesion, cohesion, and tackiness of industrial greases. This is gaining importance as new greases are formulated for high-temperature applications like in automotive, aeronautics and metal forming [13]. Furthermore, the repeatability and reproducibility of this method are discussed, as they are key criteria for the systematic development and ranking of greases. Finally, in this work, a first attempt is made to investigate the influence of grease degradation due to wear on its adhesion and ability to form threads.

2. Materials and Methods

The adhesion and tackiness of greases were measured with a Falex Tackiness Adhesion Analyser (TAA) tester (Figure 3a). As counter-material, a 3 mm Ø copper ball was used (Figure 3b). The selection of a point contact and a copper ball is based on the authors' previous experience [9], where they observed better repeatability of measurements with this combination. The reason is that point contacts create well-defined individual strings, whereas copper is a relatively inert metal. Different contact geometries (e.g., flat-on-flat) and material combinations can also be used. For this method, a specialised 316 stainless steel holder with a surface roughness (Sa) of ≤0.4 µm (ISO 25178) was developed. This holder consists of 15 individual grease scoops (Figure 3c). Each grease scoop was filled in carefully with approximately 1 mL of grease and then carefully spread with a spatula, forming a homogenous flat surface (Figure 3c). The 15 scoops can be then used to perform 15 different indentation–retraction test profiles with variable conditions (e.g., retraction speeds, contact load and temperatures).

Figure 3. (a) Tackiness Adhesion Analyser (TAA), (b) mN load sensor with 3 mm Ø Cu ball attached to it and (c) grease scoop holder with 15 individual pots (positions).

A 15-step profile was programmed to perform indentation–retraction measurements under 5 different retraction speeds and 3 temperatures in one automatic run. The complete profile is given in Table 1. Ten repeats (cycles) were performed per condition for the same grease to get information about the repeatability of the method and to perform statistical analysis of the obtained data.

Table 1. Indentation–retraction test profile.

Step	Retraction Speed (mm/s)	Temperature (°C)	Retraction Distance (mm)	Repeats (Cycles)
1	0.1	30	10	10
2	0.5	30	10	10
3	1	30	10	10
4	2	30	10	10
5	5	30	10	10
6	0.1	60	10	10
7	0.5	60	10	10
8	1	60	10	10
9	2	60	10	10
10	5	60	10	10
11	0.1	90	10	10
12	0.5	90	10	10
13	1	90	10	10
14	2	90	10	10
15	5	90	10	10

During each indentation–retraction measurement (cycle), the force on the load sensor was measured as a function of the displacement of the indenter, and the absolute values of the pull-off force, thread length and separation energy (as explained in Figure 2) were extracted and analysed with an Excel macro software and with OriginLab® OriginPro 9 software.

In this work, 8 fully formulated commercially available greases were tested. A description of these greases is given in Table 2. However, due to confidentiality, their names and compositions are not included.

Table 2. Overview of greases used in this work.

Grease	NLGI Grade	Thickener	Base Oil
A	2	Lithium complex	ISO VG 220 (kinematic viscosity of 220 cSt at 40 °C)
B	2	Lithium	Mineral base (kinematic viscosity of 220 cSt at 40 °C
C	1–2	Complex calcium sulphonate	Synthetic (PAO) mineral (kinematic viscosity of 80 cSt at 40 °C)
D	00	Paratac	Paraffinic base oil (kinematic viscosity of 90 cSt at 40 °C)
E	2	Lithium complex	Mineral base (kinematic viscosity of 150 cSt at 40 °C
F	2	Polymer-modified lithium	Mineral base (kinematic viscosity of 115 cSt at 40 °C
G	2	Polymer-modified lithium	Mineral base (kinematic viscosity of 115 cSt at 40 °C
H	2	Calcium	mineral base (kinematic viscosity of 115 cSt at 40 °C

3. Results and Discussion

3.1. Effect of Temperature

The performance of greases is strongly dependent on the contact conditions/geometry and the environment under which they operate [9]. This method can perform multiple measurements of adhesion and tackiness under varying conditions automatically so that the speed, load and temperature can be recorded with one run. Based on the authors' previous experience [6–12], an optimised test profile is proposed, as presented in Table 1. In this profile, the speed and temperature vary, whereas the load remains constant. The reason for not changing the load, too, is that for the selected contact geometry (point contact) and load range, the adhesion and tackiness appear to be less prone to load variations. Up to now, the majority of research work on adhesion and tackiness has focused mainly on contact conditions and less on the temperature effect, which is a proven factor influencing the performance and lifetime of greases [14–18]. Indeed, as can be seen in Figure 4, for the selected greases and temperature range, the increase in temperature leads to an increase in both adhesion (pull-off force) and tackiness (thread formation). However, depending on the formulation of the grease a different behaviour can be monitored. For example, the pull-off force of grease A significantly increases when increasing the temperature from 30 to 90 °C (Figure 4a). This indicates that the grease attaches more firmly onto the steel substrate (bottom of the grease scoop). In addition, this increase appears to be more gradual for grease A (Figure 4a) than for greases B (Figure 4c) and C (Figure 4e), where a sharp rise increases at 90 °C. A similar behaviour in terms of adhesion is observed for greases B and C. It is also very interesting that the same temperature dependence is observed for both retraction speeds, namely 0.1 and 1 mm/s, which is an indication that this behaviour is temperature related. Indeed, the viscous and viscoelastic responses of greases are significantly influenced by temperature changes [19–21]. In addition, phase transitions within the microstructure of the grease can also influence its yield strength [22] and force required for the deformation and separation of this layer.

When evaluating the tackiness (thread formation) of greases A, B and C (Figure 4b,d,f), a different temperature effect is observed between them, as, for example, grease C seems to be less temperature dependent (for the selected range) than greases A and B. Furthermore, it should be pointed out that adhesion (pull-off force) is a different grease characteristic than tackiness (thread length). For instance, when comparing greases A and C, grease A has a higher pull-off force than C but forms shorter threads. To put it in simple terms, grease A is stickier than grease C, yet grease C is tackier than grease A.

To illustrate that adhesion and tackiness are two different intrinsic properties of greases, a comparative test was designed and performed. In particular, a duplicate test was performed for the same grease, counter-material and test conditions, but in the second run, anti-stick paper was fixed to the bottom of the grease scoop, as shown in Figure 5a. The interesting outcome of these tests is that the thread formation (tackiness) is not affected by the use of anti-stick paper (Figure 5b), but on the contrary, the pull-off force (stickiness) drops significantly (Figure 5c). This clearly indicates that adhesion and tackiness should be considered separately.

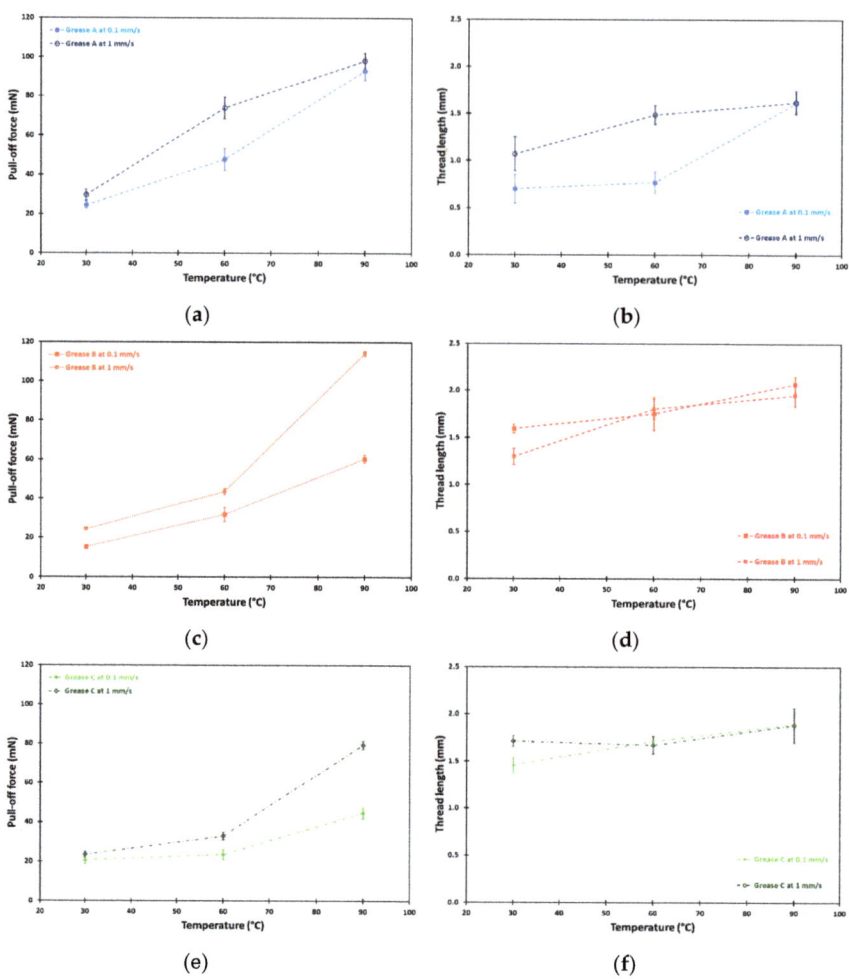

Figure 4. Effect of temperature and retraction on the pull-off force and thread formation of greases (**a**,**b**) A, (**c**,**d**) B and (**e**,**f**) C.

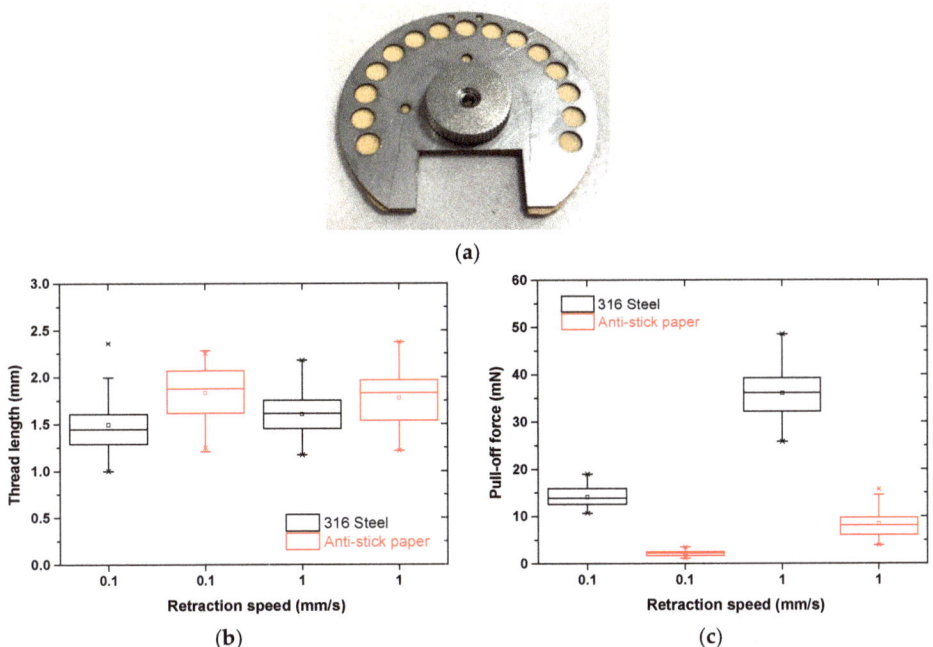

Figure 5. (**a**) Grease scoop with anti-stick paper. Effect of anti-stick paper on the (**b**) thread formation (tackiness) and (**c**) pull-off force (adhesion) of grease D for 0.1 and 1 mm/s retraction speed at 30 °C.

3.2. Repeatability and Reproducibility of Method

A significant step towards the standardisation of this method is to evaluate its repeatability and reproducibility. To achieve this, firstly the influence of the test apparatus was evaluated by testing the same grease by the same user and under the same test conditions with nine different TAA modules (see Figure 6). Results in terms of pull-off force and thread length were similar between the nine modules, as the fluctuation within one test group is in the range of 8–16% for the pull-off force and 4–12% for the thread length and between the different modules around 8% and 7%, respectively, at 0.1 mm/s. When increasing the retraction speed, the fluctuation between the individual test groups can rise up to 25% (depending on the grease). In particular, for the same grease, when the retraction speed is 1 mm/s, the fluctuation within one test group is in the range of 4–25% for the pull-off force and 6–12% for the thread length and between the different modules is around 16% and 11%, respectively.

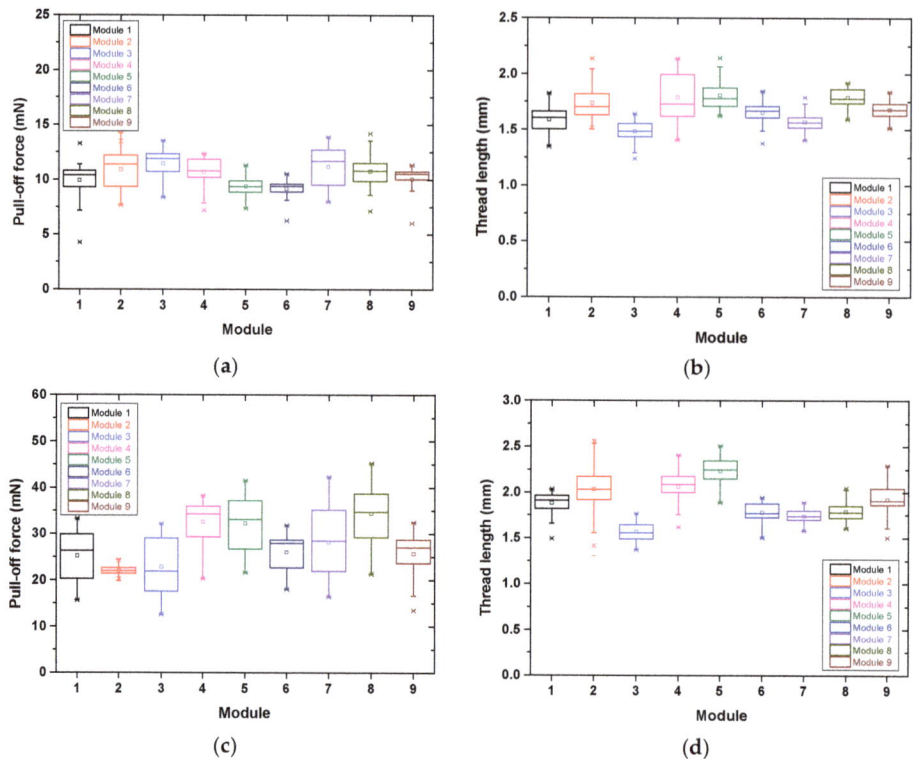

Figure 6. Comparison of 9 different modules for grease E. (**a**) Pull-off force and (**b**) thread length for 0.1 mm/s retraction speed at 30 °C. (**c**) Pull-off force and (**d**) thread length for 1 mm/s retraction speed at 30 °C.

What is also very interesting is that the repeatability of other greases, under lower speed retraction conditions (0.1 mm/s), is also in the range of 8–16% for the pull-off force and 5–12% for the thread length for each module (Figure 7). Again, the fluctuation between groups is similar to the variability between the different modules, which is around 8% for the pull-off force and 7% for the thread length.

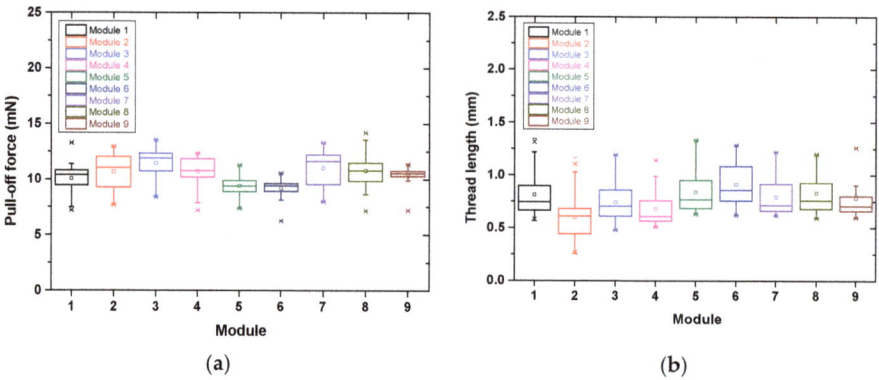

Figure 7. Comparison of 9 different modules for grease F. (**a**) Pull-off force and (**b**) thread length for 0.1 mm/s retraction speed at 30 °C.

As a second step, a comparison between users was performed for the same grease, module and test conditions, as presented in Figure 8. The fluctuation between the five users at 0.1 mm/s is 7–12% for the pull-off force and 5–13% for the thread length. These values fall again within the fluctuation observed in each measurement group (one user, one module and one grease), which is approximately 8% for the pull-off force and 6% for the thread length at 0.1 mm/s speed.

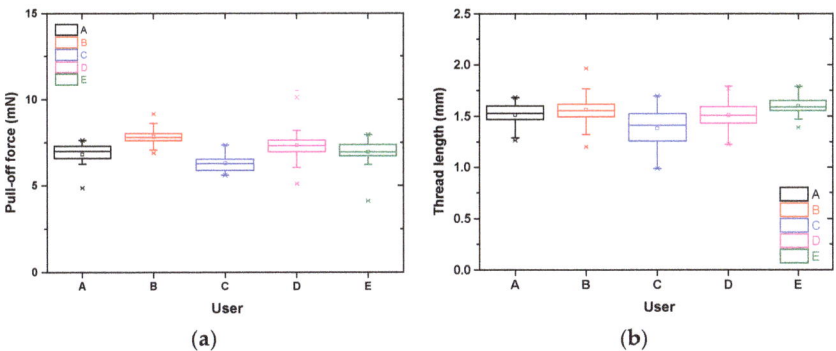

Figure 8. Comparison of (**a**) pull-off force and (**b**) thread length between 5 different users for the same modules for grease G. Test conditions: 0.1 mm/s retraction speed at 30 °C.

It should be pointed out that the reproducibility of grease measurements with other standard methods like ASTM 2509 Timken tests is 73% and X*sqrt(average) for ASTM D4170 Fretting tests. Thus, the authors believe that the developed protocol is quite repeatable and in line with other standardised methods and that the observed fluctuations are mainly due to the heterogeneous and complex nature of greases [23].

3.3. Other Possibilities: Grease Monitoring

One of the potential applications of this method is to use it to easily monitor changes in the behaviour of a grease. Since only a small quantity of grease is needed per test, it can be used to evaluate changes in the adhesion and tackiness of greases in the field, as, for example, in a wind turbine. Thus, it can be potentially used to monitor changes in the structure of a grease. For instance, the effect of wear on the adhesion and tackiness of a grease can be seen in Figures 9 and 10. Indeed, the more severe the wear conditions (point contacts in four-ball vs. line contact for Timken tests) and the smaller the quantity of the grease used (15 g four-ball vs. 5 kg for Timken tests), the more significant the effect of wear degradation on the pull-off force and thread length.

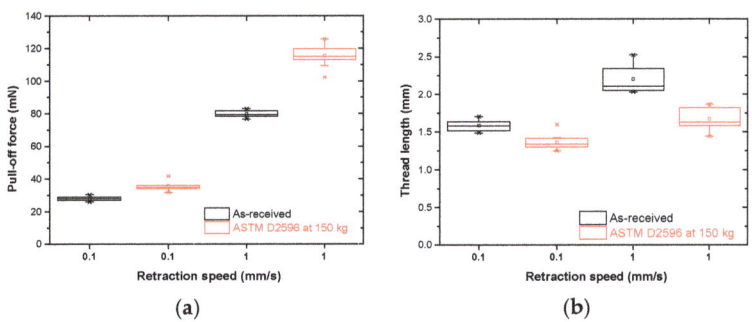

Figure 9. Effect of 4-ball wear testing (ASTM D2596) on the (**a**) pull-off force (adhesion) and (**b**) thread length (tackiness) of grease H.

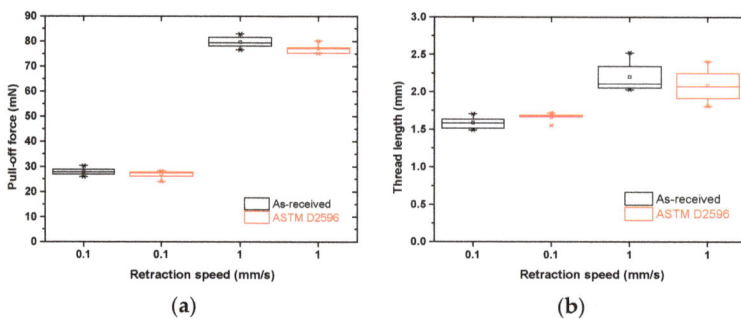

Figure 10. Effect of Timken testing (ASTM D2509) on the (**a**) pull-off force (adhesion) and (**b**) thread length (tackiness) of grease H.

4. Conclusions

In this work, the most recent progress on measuring the adhesion and tackiness of greases with the indentation–retraction approach is presented. In particular, the ability of a grease to adhere to a surface (pull-off force) and to form threads when it is being pulled apart (tackiness) strongly depend on the temperature. However, the influence of temperature is not the same for these two grease characteristics. In addition, it was proved that adhesion is strongly dependent on the holder, which means that this property is system oriented. On the other hand, thread formation does not seem to be influenced; thus, it is more of an intrinsic characteristic of the grease. As a follow-up, the influence of other environmental factors such as humidity and material composition will be investigated.

For this method to move towards standardisation, its repeatability and reproducibility should be addressed. From an internal round robin exercise, it was found that the repeatability of the method for one group of measurements (one module, one tester and one grease) is in the range of 5–15%. A similar fluctuation was found between the different modules for the same grease and user, as well as between different users for the same grease and module. From the authors' point of view, this indicates that the method is quite repeatable and reproducible and that the fluctuation is possibly due to the heterogeneous and complex grease structure. However, to draw safe conclusions, a more extended round robin is underway between different labs and well-defined and formulated greases. Another possibility of this method, apart from grease formulation and development, can also be grease condition monitoring.

Author Contributions: Tester development, O.M. and M.C.; testing, E.P.G., D.D. and M.D.B.; data analysis, E.P.G., D.D. and M.A.; writing, E.P.G. and D.D.; review, M.A. All authors have read and agreed to the published version of the manuscript.

Funding: This research received no external funding.

Institutional Review Board Statement: Not applicable.

Informed Consent Statement: Not applicable.

Data Availability Statement: Not applicable.

Conflicts of Interest: The authors declare no conflict of interest.

References

1. Bikerman, J.J. The fundamentals of tackiness and adhesion. *J. Colloid Sci.* **1946**, *2*, 163–175. [CrossRef]
2. Barquins, M.; Maugis, D. Tackiness of elastomers. *J. Adhes.* **1981**, *13*, 53–65. [CrossRef]
3. Gay, C.; Leibler, L. Theory of tackiness. *Phys. Rev. Lett.* **1998**, *82*, 936–939. [CrossRef]
4. Moon, M. On the right tack. *Lubes'n'Greases* **2018**, *24*, 64–69.
5. Morway, A.J.; Young, D.W. Grease Compositions. U.S. Patent 2,491,055, 13 December 1949.

6. Achanta, S.; Jungk, M.; Drees, D. Characterisation of cohesion, adhesion, and tackiness of lubricating greases using approach–retraction experiments. *Tribol. Int.* **2011**, *44*, 1127–1133. [CrossRef]
7. Georgiou, E.P.; Drees, D.; De Bilde, M. The quantitive method for measuring grease tackiness. *Lube Mag.* **2016**, *132*, 18–19.
8. Georgiou, E.P.; Drees, D.; De Bilde, M.; Anderson, M. Can we put a value on the adhesion and tackiness of greases? *Tribol. Lett.* **2018**, *66*, 60. [CrossRef]
9. Georgiou, E.P.; Drees, D.; De Bilde, M.; Anderson, M. Quantitative approach to measuring the adhesion and tackiness of industrial greases. *Tribol. Und Schmier.* **2019**, *66*, 37–43.
10. Harmon, M.; Powell, B.; Barlebo-Larsen, I.; Lewis, R. Development of a grease tackiness test. *Tribol. Trans.* **2019**, *62*, 207–217. [CrossRef]
11. Georgiou, E.P.; Drees, D.; De Bilde, M.; Anderson, M. How reliable and sensitive is the new Indentation/Retraction method in measuring tackiness of industrial greases? In Proceedings of the 74th STLE Annual Meeting and Exhibition, Nashville, TN, USA, 19–23 May 2019.
12. Georgiou, E.P.; Drees, D.; De Bilde, M.; Feltman, F.; Anderson, M. Grease Adhesion and Tackiness: Do They Influence Friction? *NLGI Spokesm.* **2020**, *84*, 20–25.
13. Mang, T. *Encyclopedia of Lubricants and Lubrication*; Springer: Berlin, Germany, 2014.
14. Pan, J.; Cheng, Y.; Vacca, A.; Yang, J. Effect of temperature on grease flow properties in pipes. *Tribol. Trans.* **2016**, *59*, 569–578. [CrossRef]
15. Wang, H.; Li, Y.; Zhang, G.; Wang, J. Effect of temperature on rheological properties of lithium-based magnetorheological grease. *Smart Mater. Struct.* **2019**, *28*, 035002. [CrossRef]
16. Hurley, S.; Cann, P.M.; Spikes, H.A. Thermal degradation of greases and the effect on lubrication performance. *Tribol. Ser.* **1998**, *34*, 75–83.
17. Lugt, P.M. Modern advancements in lubricating grease technology. *Tribol. Int.* **2016**, *97*, 467–477. [CrossRef]
18. Booser, E.R.; Khonsari, M.M. Grease life in ball bearings: The effect of temperatures. *Tribol. Lubr. Technol.* **2010**, *66*, 36–44.
19. Williamson, B.P.; Walters, K.; Bates, T.W.; Coy, R.C.; Milton, A.L. The viscoelastic properties of multigrade oils and their effect on journal-bearing characteristics. *J. Non-Newton. Fluidmech.* **1997**, *73*, 115–126. [CrossRef]
20. Delgado, M.A.; Valencia, C.; Sanchez, M.C.; Franco, J.M.; Callegos, C. Thermorheological behaviour of a lithium lubricating grease. *Tribol. Lett.* **2006**, *23*, 47–53. [CrossRef]
21. Delgado, M.A.; Valencia, C.; Sanchez, M.C.; Franco, J.M.; Callegos, C. Influence of soap concentration and oil viscosity on the rheology and microstructure of lubricating greases. *Ind. Eng. Chem. Res.* **2006**, *45*, 1902–1910. [CrossRef]
22. Cyriac, F.; Lugt, P.M.; Bosman, R. On a new method to determine the yield stress in lubricating grease. *Tribol. Trans.* **2015**, *58*, 1021–1030. [CrossRef]
23. Lugt, P.M.; Velickov, S.; Tripp, J.H. On the chaotic behavior of grease lubrication in rolling bearings. *Tribol. Trans.* **2009**, *52*, 581–590. [CrossRef]

MDPI
St. Alban-Anlage 66
4052 Basel
Switzerland
Tel. +41 61 683 77 34
Fax +41 61 302 89 18
www.mdpi.com

Lubricants Editorial Office
E-mail: lubricants@mdpi.com
www.mdpi.com/journal/lubricants

www.ingramcontent.com/pod-product-compliance
Lightning Source LLC
LaVergne TN
LVHW070617100526
838202LV00012B/667